SERICULTURE AND PEST MANAGEMENT

Sericulture and Pest Management

A.D. Jadhav
T V Sathe

2026
Daya Publishing House®
A Division of
Astral International Pvt. Ltd.
New Delhi – 110 002

First Published, 2012

Reprinted 2026

ISBN: 9788170352419 (HB)

Published by : **Daya Publishing House®**
A Division of
Astral International Pvt. Ltd.
– ISO 9001:2015 Certified Company –
4736/23, Ansari Road, Darya Ganj
New Delhi-110 002
Ph. 011-43549197, 23278134
E-mail: info@astralint.com
Website: www.astralint.com

Digitally Printed at : **Replika Press Pvt. Ltd.**

PREFACE

Sericulture is an agro-based cottage industry and potential source of earning foreign currency and capable of generating substantial, gainful employment in the rural areas with very little capital investment. In global silk scenario China, India, Japan, Russia, etc. are leading countries. China ranks first while India second in mulberry silk production. India has unique distinction of being the only country in the world to produce all four types of commercially known silks viz., Mulberry, Tasar, Mugs and Eri. Silk has a very large domestic market in India which is growing steadily.

The global silk scenario has changed drastically in the recent years both in silk producing and consuming countries, the demand of silk is more than one lack tons per year. There is a vast supply gap in India of around 4000 tons of raw silk. Thus, sericulture business has tremendous scope in India. The major silk production in India is from five traditional states, Karnataka, Andhra Pradesh, Tamil Nadu, West Bengal and J. & K. Many non-traditional states like Maharashtra, Madhya Pradesh, Uttar Pradesh, Meghalaya, Manipur, Nagaland etc., are emerging very fast.

Biotechnology helped a lot to sericulture however, expected yield of cocoon production is not achieved so far because of the attack of several pests and diseases to silkworms. Uzi fly, Dermestid beetles, Ants, Pentatomids, Praying mantis, Ichneumonid and Braconoid wasps are the important pests of silkworms. Among the above pests, uzi ranks first. It is known to cause more than 40 per cent damage to silkworms in some states of India. The control of various pests is a very essential fact and chronic problem. Pesticidal control has several adverse effects hence, biological control received considerable attention from the experts from the field of uzi fly management but, uzi menance is still unsolved problem.

The book contain nine chapters. The first chapter general introduction narrates national and international status of sericulture. The second chapter includes moriculture, the third disinfectants, the fourth seed production and the fifth rearing of silkworms (Mulberry and Tasar). The sixth chapter covers the survey and survelliance of natural enemies. The seventh chapter embodies the aspects of parasitism, life history, rearing host age selection, host specificity and life table statistics and intrinsic rate of increase. The eighth chapter represents the summary of the text and the ninth bibliography referred.

The book explores several avenues for future research in the field of sericulture. The aspects covered would be useful for better rearing of silkworms and designing control measures of natural enemies of silkworm and making the host parasitoid index of the region. We feel that the book will find interesting and stimulatory to research students, teachers, farmers and sericulturists.

T.V. Sathe
Kolhapur **A.D. Jadhav**

CONTENTS

Chapter 1
INTRODUCTION

The vision of future India will only depend on the recognition of "Science and technology" and its proper application, that we can initiate sustainable integrated development to ensure food, economics and health security to all the people in the country (Rao, 1996). Over the last two hundred years, total cultivated land area has increased from 600 million ha. to 1500 million ha on a global basis. Even though estimates indicate that the cultivable land across the globe can be increased from the present 1500 million ha. to about 2100 million ha through reclamation of degraded and wasteland, the possibility of addition to arable land in Asia including India is less than 10 per cent (Rao, 1996). However, land for cultivation for mulberry, have not increased significantly in India. The area under irrigation in India has also correspondingly increased from less than 20 million ha in 1950 to about 50 million ha in the last four decades.

In India, sericulture is one of the important agro and forest based cottage industry, earning a foreign exchange of Rs. 400 crores/annum and providing gainful employment to over six million people. Technically, sericulture is rearing of silkworms either on mulberry or non-mulberry plants for production of silk.

Biotechnology can be employed for improving the productivity and quality of Indian silk, host-plant improvement and product utilization. However, in sericulture, impacts of these techniques have yet to be felt.

There are major areas in which the modern methods of biotechnology find a place in sericulture. For development of better races of silkworm, for production of other commercially important biomolecular products and also for studying basic biology, genetics and developmental biology, Department of Biotechnology has launched a programme which has short medium and long term projects which will concentrate on R & D work at laboratory level and large scale demonstration at field level.

The high rate of soil erosion in deforested areas in our country ranges from 10 tons/ha. in th plains to about 30 tons/ha. in the North-Eastern hilly region, ten to fifty times larger than in the forested areas. The resulting annual sedimentation load from Ganga and Bramhaputra alone exceed 1500 million tons, making them the two largest dumpers of suspended load anywhere in the world. The environmental problems caused by increased sedimentation has extensively affected the capacity of reservoirs as well as the aquatic ecosystem in the inland waters just as soil erosion, in addition to loss of fertility, has drastically reduced the recharging capacity, resulting in sever depletion of underground water level. Extensive deforestation has resulted in increased carbon dioxide concentration in the atmosphere, increased albedo from deforested areas altering the energy balance between the terra-ferma and the atmosphere, increased run off of rain precipitation from 20 per cent to almost 50 per cent and a gradual extinction of bio-diversity. The soil erosion and flora and fauna can be protected by cultivation of mulberry.

Rapid deforestation over the years due to the pressure of population and increased industrialisation have reduced our total forest cover to just 20 per cent of the land area, only half of which is closed forest over 40 per cent canopy. These figures must be viewed against the background of the National Policy adopted in 1952, recommending

33 per cent forest cover across the country with 60 per cent coverage in hilly areas (Rao, 1996). An inevitable consequence of deforestation is increased run off rain water precipitation and extensive soil erosion, resulting in unacceptable deterioration of top soil, degradation of top soil, degradation of land and sedimentation of water bases resulting in desertification and frequent flooding. It is absolutely essential to inject the environmental concern into the whole process of national development. Over the past few years, a more comprehensive view of environment has grown which brought within the ambit of our environmental concerns, the broader issues are relating to resource management, pollution and waste management, and human settlements. The resource management would cover depletion of non-renewable resources, fossil fuel, degradation and over consumption of renewable resources, biomass and soil, destruction of genetic resources, desertification, deforestation etc. Sericulture in deforestation and other areas would certainly add relevance in improving the situation for erosion and deforestation by planting Mulberry (Plate II, Fig. 11), Ain, Arjun (Plate II, Fig. 12) etc. as a food source of silkworms.

Significant increase in rural income and employment opportunities is of almost importance for improving the quality of life in rural areas and to check the migration of people to urban conglomerates. The absurd concept of rural economy, based essentially on the sale of raw agricultural products has led to the deplorable situation of the per capita income of the rural poor continuing to remain just one-fifth of that of his urban counterpart, in spite of the country gaining self-sufficiency in food. Sericulture provides tremendous opportunity for employment to the rural landless people and to check migration. It also provides opportunity for utilising the waste and dry lands and provides handsome money to the poor rural marginal farmers (land holdings of ½ acre or so). When nothing could be grown in a depleted land, sericulture may found successful.

According to natural forest policy, 33 per cent of the land should be covered with forest to maintain the ecosystem. India has only 23 per cent of the total land area under forests. Recent remote sensing aerial photography has shown that the percentage declined to 12 per cent. This shrinkage in forest area is due to the onslaughts of firewood gatherers, land hungry farmers and industrial timber interests. Rapid population growth is another factor causing the depletion of forest, reminding one of the remark. "The forest came before the civilization and deserts after it."

At the present rate of the assault on forests, India is steadily losing more than 1.5 million ha, out of 11 million ha. of world's tropical forests land lost each year. In other words, once every five years forest cover of the size of the state of Haryana vanishes and over three decades, would equal an area of about the size of India. Hence, social forestry programme is the need of the hour.

The term sericulture is not only related to rearing and production of silk from mulberry silkworms, *B. mori* but also from other non mulberry silkworms. Plantation of tasar food plants, for tasar cultivation as well as mulberry in fallow/barren land will certainly strengthen the bulk of green area up to some extent. Tasar culture can act as a harmonious linkage between tribal welfare and forest conservation and can translate the social forestry programme into reality (Siddiqui, 1988).

Mulberry is tolerant and cultivable to a wide range of agro-climatic conditions. The agro-forestry is a sustainable land management system in which a woody perennial is raised in conjunction with arable crops and/or live stock, either in sequence or simultaneously on the same piece of land. It has been classified from the stand point as agro-ecological zone and socio-economic aspects. The major systems of agroforestry reported by Nair (1988) are as follows :

1. Shifting cultivation,
2. Taungya cultivation,
3. Home-gardening or multitier system,
4. Agro-Sericulture system,
5. Sericulture system,
6. Silvi Pastoral system,
7. Alley cropping or hedge row system,
8. Agri-Silvi horticulture and
9. Wind break and shelter belt.

Mulberry has a lot of feasibility and possibility to overcome under these systems (Singh and Ghosh, 1992).

Sericulture is an admixture of forestry, agriculture, industry and art. The underlying philosophy of tasar culture encompassed all the important areas like forest production, organized systematic use of resources and environment, self-employment, proliferation of industries and trade, promoting education as well as academic and applied research. Tasar culture/mulberry plantation can also become one of the most effective way implementing the main themes of new 20-point programme eradicating poverty raising productivity, reducing income inequalities, and remaining social and economic disparities and improving the quality of life, etc. (Siddiqui, 1988).

Sericulture is well-known as a highly employment oriented and a low capital intensive activity ideally suited to the conditions of an agro-based industry. It is an effective tool for rural development with overwhelming number of advantages. It transfers wealth from richer sections of society to poorer since, silk is consumed mostly by the affluent people and the money so spent on purchases of silk is distributed among the sericulturists, reelers, twisters weavers and traders, with 51.5 per cent, 6.2 per cent, 8.2 per cent, 14.5 per cent and 19.5 per cent respectively (Nair, 1988). Sericulture, the agro-based cottage industry, employing 51.62 lakh persons mostly in the rural areas. Thus, it prevents migration of rural population to urban. It provides highest employment opportunities (25.23 lakh persons per year) as compared to other occupations. It creates employment at all the stages viz. mulberry cultivation (5.20 lakh), silkworm rearing (5.83 lakh), reeling (0.71 lakh) and weaving (7.23 lakh) and viewed as one of the best alternative sources of employment.

The role of tasar in sericulture and industrial development is now well established as an ideal forest based industry which can meet the needs of our socio-economic conditions. Tasar rearers mostly belongs to various tribes and are the weaker sections of our society, mostly living in remote forests. The tribals account for nearly 8.10 per cent of country's population (Census, 1991). Out of the total tribal population of 38 million, over 97 per cent are confined to the rural areas mostly inhabiting the forest belts. Nearly 30 million (77%) are located in tasar cultivating states of tropical (25.67 million) and temperate (4.23 million) tasar belts (Census, 1991).

More than 13.2 lakh tribals are engaged in tasar culture in tropical zone and only five to seven thousand in temperate zone. Thus, a huge quantity of human resource still remain untrapped (99%) as far as the tasar culture is concerned. Almost an equal number is engaged in the reeling, spinning, weaving and processing of tasar cocoon and silk.

The tasar belt covers the states of Bihar, Madhya Pradesh, Orissa, Maharashtra and touching the fingers of Andhra Pradesh, Karnataka, West Bengal and Uttar Pradesh. In all 15 states are engaged in the production of non-mulberry silk. Amongst these states Bihar is the largest producer of tasar cocoons. Madhya Pradesh is next in the series so far as cocoon production is concerned and Orissa ranks third. Madhya Pradesh has 40 lakh tribal population in tasar cultivating districts and the area under tasar food plants is also vast (50.44 lakh ha). The rearing is practised in Eastern and Southern districts of Bilaspur, Raigarh, Sargunja, Balghat, Mandala and Bastar. The district Bilaspur is one of the traditional tasar districts of Madhya Pradesh with rich tribal population and forests (Siddiqui and Omkar, 1987). Sericulture is playing a complementary role in the settlement of the soligas, the semi-nomadic tribal community in B.R. Hills of Mysore district in Karnataka (Raju *et al.*, 1995). The colonization of tribal population is possible through tasar sericulture practices. This will prevent the population load on the urban area. The statewise tribal population is shown in Table 1.

Sericulture has been identified as an occupation of low investments, high output source of employment and income. The industry holds high promise as an employment intensive occupation specially in rural and semi-urban areas (Benchamin and Jolly, 1987). Un-employment and under-employment continue to be the major problems of our country, it is more serious in rural areas. Out of 306 million people living below poverty line in India, 249 million are in rural areas and only 57 million in urban centres. Seventy per cent of the total workforce still depend upon agriculture sector for their livelihood and this figure has remained more or less constant during the past two decades despite considerable industrialization. In labours, surplus economy and worthwhile planning strategy should have maximised for employment opportunities as one for of its key objectives. It is here that Sericulture industry fits into our socio-economic condition as a tool for rural development (Benchamin and Jolly, 1987). Out of the 5.76 lakh villages in the country, sericulture is practised in about 50,000 villages, providing employment to more than 60 lakh people in rural and semi-urban areas of whom more then 30 per cent are from the backward sections of the society and tribals. It is estimated that one acre of mulberry provides gainful employment opportunities indirectly in ancillary units like manufacture of equipments and marketing of sericultural products.

Table 1: Statewise Scheduled Tribe Population in India (1991 Census)

State/Union Teritory	Total tribal population	ST population	% ST population
Andhra Pradesh	66,508,008	4,199,481	6.31
Arunachal Pradesh	864,558	550,351	63.66
Assam	22,414,322	2,874,441	12.02
Bihar	86,374,465	6,616,914	7.66
Goa	1,169,793	376	0.03
Gujarat	41,309,582	6,161,775	14.92
Haryana	16,463,648	—	—
Himachal Pradesh	5,170,577	218,349	4.22
Jammu & Kashmir	7,718,700	—	—
Karnataka	44,977,201	1,915,691	4.26
Kerala	29,098,518	320,967	1.10
Madhya Pradesh	66,181,170	15,399,036	23.27
Maharashtra	78,937,187	7,318,281	9.27
Manipur	1,837,149	632,123	34.41
Meghalaya	1,774,778	1,517,927	85.53
Mizoram	689,756	653,565	94.75
Nagaland	1,209,546	1,060,822	87.70
Orissa	31,659,736	7,032,214	22.21
Punjab	20,281,969	—	—
Rajasthan	44,005,990	5,474,881	12.46
Sikkim	369,451	90,901	22.36
Tamil Nadu	55,858,956	574,194	1.03
Tripura	2,757,205	853,345	30.95
Uttar Pradesh	139,112,287	287,901	0.21
West Bengal	68,077,965	3,808,760	5.59
Union Territories			
A & N Island	280,661	26,770	9.54
Chandigarh	642,015	—	—
D & N Haveli	138,477	1,09,380	78.99
Daman & Diu.			
Lakshdweep	51,707	48,163	93.15
Pondichery	907,785	—	—
Total	846,302,688	67,758,380	8.01

Rural economy in India is largely dependent on agriculture. There is a considerable scope to increase employment and income in rural areas through labour intensive remunerative enterprises like sericulture. Sericulture industry is unique in its advantages and suitability to the rural setup.

In upliftment of women life, sericulture can play a very crucial role by providing the job opportunities. A large part of rural women population is underemployment in India. Women constitute about 48 per cent labour of total population.

The rural women needs employment as they cannot fulfil daily requirement with the traditional job. Even if a family has some land, the women working on the others field to supplement their family income. Yet, the women has very little say regarding the use of her earnings or any decision regarding where she is to work or how much she is to be paid. Due to mechanization, women labour is getting further pushed into the unskilled category. This entire process ensures the continued marginalisation of women. If sericulture is to be taken up as a true developmental factor for women then this industry can provide an income generating activity which has multiple advantages for women, (1) She can do in the home, women gets a fairer deal from the economy for less effort, she earns more. (2) She can combine this work with her other duties. (3) She can control therefore, her timing, rather than having to be at some other person's beck and call. (4) She can control her own earnings, the harder she works, more she earns. (5) She is ensured of a year-round income. (6) She learns to deal with people outside her home/community/village and develops her won personality and self-confidence.

Sericulture is a sustainable economic activity positively favouring the rural poor in the unorganised sector. An estimated 60 per cent of the worker force engaged in sericulture is said to be women. Rural women sericulturists have to be strengthened as producers to help them to increase productivity and returns. In this, instance women have to be reached with credit, access to resources such as land, inputs, technology, training, information as well as the materials with which they can become sustainable producers (Venugopal, 1994). It is possible to women to supplement the other business like intercropping and introducing stall fed cattle, sheep, rabbits and open fed poultry for additional income from the parental crop mulberry. It is commonly recognised that women, in all traditional sericultural states, constitutes a substantial proportion (53%) of the total work force in sericulture and contribute vitally to the sectors of silkworm rearing, reeling of cocoons and even in twisting, weaving, printing and dying (Gayathri Devi, 1994). At present more than 75,000 women are actively involved in sericulture activity in India.

Sericulture is agrobased industry in which mulberry and non-mulberry food plants are used for rearing the different type of silkworms for silk production, reeling and further for fabrication.

India is the unique country in the world where all the four varieties of silk namely Mulberry, Tasar, Eri and Muga are produced. The silkworm is a common name for the silk producing group of Lepidopteran insects, they are grouped on the basis of their feeding habits and domestication. Silk produced by the silkworm is named differently on the basis of feeding host plant. Silkworms belong to super family *Bombycoidea* which comprises as eight families from which two families are economically important those are *Bombycidae* and *Saturniidae*. The members of these families which produce commercial value silk are given below :

1. *Bombyx mori* Linn. — *Bombycidae*
Domesticated mulberry silkworm

2. *Bombyx mandarina* — *Bombycidae*
Wild ancestor of mulberry silkworm

1. *Antheraea pernyi* G.M. — *Saturniidae*
It is a Chinese tasar silkworm

2. *Antheraea mylitta* Drury — *Saturniidae*
It is a Indian tasar silkworm

3. *Antheraea yamamai* G.M. — *Saturniidae*
The Japanese tasar silkworm

4. *Philosamia cynthia* D. — *Saturniidae*
A wild species of Eri silkworm

5. *Philosamia ricini* Boisd. — *Saturniidae*
The eri silkworm

6. *Eriogyna pyretorum* — *Saturniidae*
Fishline silkworm Hainan Island

Among the four types of silk produced in India the mulberry silk is commercially the most important and it constitutes about 95 per cent of the world production and is a most commonly known and commercially available silk in the market.

Mulberry silkwork *Bombyx mori* Linn. is reared on mulberry plants. A very large number of mulberry varieties are known to science on which quality cocoons can be produced. Mulberry can be grown in temperate and tropical zones of the world with varied environmental conditions. More than 60 countries are involved in cultivation of mulberry for sericulture business.

Mulberry is medium sized deciduous plant belonging to family Moraceae. More than one thousand varieties have been reported from Japan. While 798 have been reported from India (Dandin and Ramesh, 1987). In India, the area under mulberry cultivation is 2,27,586 ha (Dandin and Ramesh, 1987) and 2,16,734 ha mulberry is under irrigation in South India. Rain fall in the arid and semi-arid zones is very erratic and the prevailing drought conditions adversely affect mulberry leaf production. Karnataka which has an annual rainfall of 500-700 mm, is a main centre for the production of cocoons in India (Ramakant, 1998).

China stands number one in the world silk production, which accounts for two-thirds of total cocoon and silk production. Most parts of China receive moderately high rainfall and the temperature ranges between 10-20 °C. Mulberry is grown as monocrop and to some extent as border or tree. Mulberry is mainly propagated through grafts which is carried out during March to April stock. Mulberry is generally resistant to pest and diseases and also adverse conditions. China has rich mulberry resource. Out of 26 species reported, 14 species are distributed with more than 2000 farm plasma varieties. It currently has about 40 good commercial mulberry varieties adopted to various areas. Husang 32, Husang 197 and 7, the main varieties, belongs to *Morus latifolia*. Recently, evolved varieties like Yu2, Yu237, are also quite popular. The promising mulberry varieties like gloody big flower are common tree (border) varieties grown in Sichuam province. Varieties have been evolved to suit different agroclimatic conditions. Such as lunjiao belongs to *M. atropulparea* southern sub-tropics (Pearl river area) cooks comb Lv mulberry and black Lu mulberry for Yellow river area like Shandong province and heigelu and bolo mulberry varieties for Shanxi and Heben province.

In India, region specific mulberry varieties need to be developed (Govindaiah *et al.*, 1997). Mulberry, a member of family 'Moraceae' is recognized under the genus *Morus* Linnaeus as early as in 1753. Broadly, the mulberry is classified as white mulberry (*M. alba*), black mulberry (*M. nigra*) red mulberry (*M. rubra*), Indian mulberry (*M. indica*) and Russian mulberry (*M. tatarica*). Presently, it is classified into 30 species with numerous varieties. In India, along with *M. nigra* L and *M. alba,* the wild species such as *M. alba,* var. *indica* (L) Bur. (*M. indica* L.), *M. serrata* Roxb. and *M. macroura* Mig. (*M. lavigata* Wall.ex Brandis) are popular. The Indian wild species are confined to the lower Himalayan belt up to an elevation of 9,000 feet (MSL) from Kashmir to Assam. Oldest known mulberry tree is located at Joshimath (U.P.) on the way to Badrinath in the Himalayas. This tree is of 1400 years old and has a girth of 28 feet and belongs to the species *M. serrata*. Shri Adishankaracharya is known to have mediated under this tree and later on established one of the four maths of the *Adwita Cult* at this place. Hence, it is a religious and sacred tree. In the Kumaon and Sivalik hills numerous natural varieties of mulberry are found as a part of the forest flora. Himalayan belt can be considered as the centre of the origion of mulberry.

In the mulberry, sex was first demonstrated by Camararius in 1685 which laid foundation for plant breeding. Surprisingly, the sex expression in mulberry has still remained a 'puzzle' for scientists. For the purpose of mulberry improvement programme for silk industry, the wild varieties and the hybrid make a stock. For this purpose, the Japanese have accumulated as many as 1,250 types. Out of about 80 species distributed world wide, the genus *Morus* represented in India by four species and several varieties. In India presently, the collection is around 798 types, of which 320 are represented at Central Sericultural Research and Training Institute, Mysore (Dandin and Ramesh, 1987). In Maharashtra M-5 (Plate I, Fig. 2 and 3) and S-54 (Plate I, Fig. 4 and 8) are in practice. Sometimes local varieties of mulberry (Fig. 5) are also tried.

Popular mulberry cultivars in leading sericultural countries are shown in Table 2 and the geographical distribution of some important species of the genus *Morus* are recorded in the Table 3. In several states of India including Maharashtra, mulberry is cultivated in row system and paired row system (Fig. 6 and 7).

Table 2 : Popular Mulberry Cultivars in Leading Sericultural Countries

	Country	Mulberry variety
1.	India (North-East & Eastern)	Mandalya, Kejali, Berhampore S799, BC 259, TR 10.
	(South India)	Kanva-2, Mysore local, Selection 54, MR-2, V 1.
	(North India)	Sujanpur - 5, Goshoerami.
2.	China	Zung, Lurujiang-40, Tunjiang, Ching-518, Tong, Shinaganana, Tarantan Hege, Husang-7.
3.	Japan	Ichinose, Kairyoneusumigeshi (KNG), Shin - Ichinose, Kenmochi, Kosen, Goshoerami, Roso, Vasamidori.

Table 3 : Geographical Distribution of Important Species of the Genus *Morus*

Species	Country
Morus alba L.	North-Eastern China, Korea.
M. alba var. *indica* (L.) Bur. (*M. indica* L.)	India, Burma.
M. australis Poir.	Australia, India, Indonesia, Philippines, South China.
M. bombycis Koidz	China, Japan, Korea, Southern Sakhaline.
M. chinensis L.	China.
M. cathyama Hemsl.	Central China.
M. latifolia Poir.	China, Japan, South Korea.
M. macroura Mig. (*M. laevigata* Wall. ex Brandis)	India (Himalayas)
M. nigra L.	Cauacasious, Parsia, West Asia.
M. rubra L.	North America.
M. serrata Roxb.	India (Himalayas).
M. latifolia Makino	Japan, Korea.

Mulberry is the only food for rearing of *B. mori* hence, mulberry cultivation forms the basis for sericulture and plays a vital role in determining the profit. Leaf yield and quality determines the quantum, a laying which can be reared and cocoons harvested.

Mulberry has several other uses too. This plant mainly associated with silk is also useful as a fodder source, firewood, timber for building houses and boats, for sport goods, furniture, fancy items and as medicine. It also bears edible fruits. Mulberry contains different chemicals of medicinal value in leaf, fruit, stem, seed and root portions. The chemical substances are normally extracted into aqueous media in the form of decoctions and concoctions.

Mulberry leaf juice has special quality of moisturising and keeps the skin smooth and healthy, prevents throat infections, irritation and inflammation in the throat, useful in preventing high fever as febrifuge in endemic malaria. It has property of preventing diarrhoea and cold, used in the preparation of syrup which possesses refrigerant and laxative properties and laxative for infants, is good for diabeties and high blood pressure.

As like pea, cucumber and drum stick etc. mulberry leaves are also now consumed as vegetable by man. The green leaves of various plants are more nutritive, mulberry leaves, satisfy the dietary requirement of human beings and contain all the elements required for the body. In mulberry leaves, vitamin A, B, C, and D are in sufficient quantity. However vitamin C is rich. The growing mulberry leaves contain 6 to 6.8 per cent crude protein, 4 per cent soluble carbohydrates and 0.6 per cent crude fat and also several compounds of calcium, phosphorus, iron, copper, zinc etc. To produce 3000 calories of energy the recommended daily diet is 70 gram of protein, 0.8 gram of calcium, 10 mg of iron, 500 μ of vitamin A, 1.6 mg of B1, 1.8 mg of B2, 75 mg of C and 400 μ of D (Sreekumar *et al.*, 1994).

Mulberry waste leaves are quite useful as fodder. Leaf contains crude protein-13.53, crude fat 3.53, crude fibre

13.73 μ. Cattle feed on mulberry and left over shoots have been found to improve their milk yield. As mulberry leaves are valuable source of vitamins and improve the health of poultry birds and increase egg production. Chlorophyll extracted from leaf is used for cosmetics viz. soap, hair oil and also tooth paste and medicine.

Stem is useful as fencing material, medium grade fuel wood, thin and soft twigs can be used for basket like handicrafts, dry and waste shoots are used in compost. Aqueous and alkali extracts of stem and leaves are active against gram positive bacteria and yeasts. Stem bark, stripped from waste branches are used for paper pulp especially in China and Europe. The bark is used as purgative and vermifuge.

The mulberry wood is valued mainly for the manufacture of sport goods on account of its elasticity and flexibility. It is compared with teak because of shock resistant ability, strength, hardness, etc. The timber is used to produce hockey sticks, tennis rackets, presses, cricket stumps, home building, agricultural implements, furniture, tool handles, poles, shafts and bent parts of carriage and carts. Wood is also used for small boats and large ship. Mulberry wood contains tannin, extracts prepared from wood (tannin content 0.32%) which is suitable for tanning and colouring purpose.

The seed contains 25-35 per cent of a yellow dying oil with sp. gr. 150.924 to 0.926. Liquid fatty acids constitute the major part (80-90%) of the total fatty acids of the oil. Yellow dye is obtained from the roots and used in dying process. The bark of root is bitter and possess cathartic, antihelmintic and strigent properties. It is also used as restorative tonic and as remedy for nervous disorders in China.

An alkaloid deoxyjirimycin (DNJ) has been extracted from the root bark of the black mulberry, *M. nigra,* which resembles glucose and interferes with the synthesis of sugar chains. The chemical hinders the addition of sugar molecules to the outer coat of the human immunodeficiency virus. It has been observed that the surface glycoprotein of HIV is rich in sugar chain and is involved in the budding activity of the virus on and from a cell. Although the alkaloid's (DNJ's) activity at the cellular level is yet to be analysed, it is crystal clear that DNJ inhibits the enzymes glycosidases, responsible for trimming the sugar chains on glycoproteins. In a sense, DNJ hinders the docking of HIV on the human cell and hence makes the attack futile. Thus DNJ, easy to synthesize and seems to be a promise as a potential medicine against AIDS (Ray, 1989). The GD Searle Pharmaceuticals Company has taken keen interest in DNJ and is working closely with chemists at Oxford and Cambridge. Butyl-DNJ is reported more active in treating AIDS, (Ray, 1989, Tewary and Rao, 1990).

In India, 15 states are engaged in the production of non-mulberry silk. India is second largest country after China and earns about 15-40 crores of foreign exchange per annum. Tasar is practised by about 1,40,000 families in India (Krishna Rao *et al.*, 1998). The estimated work distribution in tasar sector in 1988 was, cocoon producer 42 per cent, reeler, twister and weaver 20 per cent, and trader 38 per cent (Jolly, 1998).

Non-mulberry sericulture is an age old tradition in India, practised mainly by the tribal people. It provides moderate earnings in different lean seasons of the year, when tribals do not have any work in agriculture and other allied pursuits. India is the only country bestowed by the nature to produce all the three varieties of non-mulberry silkworms and their host plants (Table 4). Of the total raw silk production of 14035 mt. (1997-98 provisional), the share of non-mulberry production, Eri, Tasar and Muga silks account for 69.78 per cent, 24.85 per cent and 5.37 per cent respectively (Shetty and Samson, 1998).

Table 4 : Non-mulberry Silkworms and Their Host Plants

Silkworm variety	Zoological name of silkworm	Food plants	
		Common name	Botanical name
Tropical tassar	*Antheraea mylitta*	Arjun	*Terminalia arjuna*
		Asan,	*T. tomentosa*
		Sal	*Shorea robusta*
		Ber	*Zizyphus mauritiana*
Oak tasar	*Antheraea proylei*	Uyung	*Quercus serrata*
	A. roylei		*Q. incana*
			Q. serricarpifolia
Eri	*Philosamia ricini*	Kesseru	*Heteropanax fragrans*
		Caster	*Ricinus cummunis*
Muga	*Antheraea assama*	Som	*Machilus bombycina*
		Sualu	*Litsaea polyntha*

Generally, people think that silk refers only to *Bombyx* Silk whose insects are reared on mulberry leaves and is a high priced, valuable and delicate textile, called the "Queen of fibres". During the past century, the silk industry has greatly contributed to the foreign exchange earnings. Among the various developing countries, silk still offers the economic prosperity. On the other hand, wild silks are not very popular, although Chinese tasar (*A. pernii*), Indian tasar (*A. mylitta*), Eri silk (*P. ricini*), Muga silk (*A. assama*) and Tensan (*A. yamamai*) have long been used for characteristic silk textiles, forming a small segment of the market (Hiromu, 1998).

Recent research on wild silks indicated that these materials possess some physiological significance in activities like controlling cholesterol in hemolymph, antibacterial functions and in UV absorption effect. These newly recognized properties may greatly increase the utilization of wild silks in the next century. The International Society for Wild Silkworms (ISWSM),

Japan has initiated research on wild silkmoths and silk since its establishment during 1980 and used available technology to harness various means of wild silk utilization. (Hiromu, 1998).

The people living in the forests and adjacent rural areas are highly dependent on the plants and their products for their daily needs including food and medicine. Non-mulberry silkworm food plants have a great medicinal value. Survey of medicinal Tasar flora in the Bhandara District was conducted with respect to their use against various diseases by the people (Yadav *et al.*, 1997). The main food plants of tasar silkworm (*A. mylitta*) which have commercial interest are *Terminalia arjuna* Bedd (Fig. 12), *T. tomentosa* W. and A., *Shorea robusta* Gaerrtn and *Zizyphus mauritiana* Lamk. The knowledge of medicinal plants is species specific and restricted to a few individuals which differs from ancient system and due to lack of written documentation, many prime remedies are vanishing with the generation. The Tasar culture would be helpful or encouraging for making index of medicinal plants of the region. Some primary and secondary food plants of tasar silkworm are useful as a medicine. The index of medicinal plants in Bhandara region is tabulated in Table 4a.

As like mulberry, the non-mulberry silkworm food plants have also other uses like furniture, paper industry, wood, etc.

Sericulture comprises of the following activities till the fabric stage :

1. Silkworm egg production,
2. Mulberry cultivation,
3. Silkworm rearing and cocoon production,
4. Reeling of raw-silk yarn,
5. Throwing of yarn or yarn preparatory activities,

Table 4a: Tasar Flora and Their Ethnobotanical Importance

Sr. No.	Family	Scientific Name	Local Name	Local medicinal uses
1.	Combretaceae	*Terminalia arjuna* Bedd.	Arjan or Arjun	Young leaves juice mixed with goat milk relieves intestinal pain. Bark useful for heart problems in powder form. Bark of tree and leaves useful as dye.
		T. bellerica Roxb.	Bhaira or Behera	*Triphalachurana* used for constipation includes ingredient of fruit, also useful in cough, asthama and other gastro-intestinal problems.
		T. chebula	Harra	Main ingredients of *triphala* to check constipation.
2.	Dipterocarpaceae	*Shorea robusta* Gaerrtn	(Sal)	Bark gum of plant is taken with candy to check dysentery.
		Ricinus communis L.	Arandi or Erandi	Seed oil is purgative and is used in reheumatism, sora is mixed with castor oil and the mixture is rubbed on the forehead to check the headache. Seeds of *Derris indica* are burnt, powdered and mixed with castor oil for treatment of skin diseases.
3.	Rhamnaceae	*Zizyphus mauritiana* Lamk	Ber	The bark extract mixed with milk and honey is used in blood purification, mild boiled bark extract is used in diarrhoea. Dried fruits are boiled with water and the filterate is used with curd for diarrhoea.

6. Yarn dyeing and
7. Weaving, printing, marketing.

The major raw-silk producing countries are China, India, Japan, CIS block and Brazil. Sericulture is also practised in other Asian countries like Thailand, Malaysia and Vietnam. China accounted 63 per cent (55,000 tons) of the world silk production (76,761 tons) in 1991.

In recent times, with modernization of equipment and increasing demand for quality and productivity, large scale farming or corporate involvement has entered in sericulture. Different approaches to this can be seen in Japan, China, Thailand, Malaysia, Brazil, South Korea, Vietnam and Indonesia.

The Japanese sericulture technology is considered as one of the best in the world. In recent years, though it has shrunk, the sericulture technologies still remain high and could be beneficial to many developing countries.

In Japan, silkworm eggs are produced by private grainages, young age rearing is conducted under controlled environment in co-operative rearing houses and chawki mechanised rearing is the specialized aspect of rearing. Late age worms are supplied to farmers for rearing in individual rearing shades. Co-operatives have tie-ups with reeling factories for marketing of cocoons. Corporate houses are also involved in supplying sericulture know-how, machinery and export of silkworm eggs.

Japan has sericulture farms in 40 per cent of its municipalities. In many districts, cocoon is an important farm product. The country's climate as well as the hard working nature and dexterity of the Japanese farmers are well suited to sericulture, and not surprisingly, silkworm rearing has long been the country's characteristic and seen considerable improvements in its technology. Unlike paddy and other agricultural crops, sericulture requires longer working period and makes it difficult for silkworm rearers to do other job. In Japan, farmers with annual income of less than five million yen generally earn additional income from non-farm jobs to continue farm management and thus, part-time farmers are on the increase, but the number of sericulturists is showing a downward trend.

In China, silkworm eggs are produced by government grainages and distributed to government agencies who in turn sell them to large government owned reeling factories. Sericultural machinery industry is government owned. China exports silkworm eggs, cocoons, raw silk yarn, fabrics and ready made garments through government agencies.

In Thailand, silkworm eggs are produced by the government and private grainages, also imported from Japan, Taiwan and China. Reeling factories conduct chawki rearing and provides late age worms for rearing to farmers with buy back arrangement of cocoons. In some cases, mulberry farms belongs to the reeling factory are given to farmers for cultivation and rearing of silkworms. Thailand exports ready made garments, cocoons and raw-silk yarn.

In Brazil, production of silkworm eggs is mainly by private grainages. Young age rearing is on co-operative basis under controlled conditions and mechanized. Late age rearing is conducted by individual farmers and cocoon purchase is by large reeling factories. Brazil exports cocoons and raw silk yarn.

In Indonesia silkworm eggs are imported from Japan. For demonstration, of mulberry cultivation, silkworm rearing units are constructed to prove the technology to local farmers. There is joint participation of the government and the corporate sector. Cocoons are produced by individual farmers or corporate farms of 300-500 acre.

Malaysia import silkworm eggs from Japan. Activation of mulberry, rearing of silkworm and reeling are under one roof. Most of projects are government owned, Malaysia exports cocoons. In South Korea, large reeling factories receive cocoons from farmers. South Korea export raw silk yarn, sericulture machinery and project know-how. Vietnam is emerging as a silk yarn producer and exporter due to government support. Government owned reeling factories buy cocoons from farmers in Vietnam.

Sericulture in the Philippines has the recent origin. Though, the beginning was made prior to the 2nd World War, the industry could not survive due to lack of technical and financial support. The Government of Philippines is having a few aid schemes to encourage the people to take up the sericulture. The farmers are getting technical and financial assistance from Philippines Textile Research Institute (PTRI) and Ministry of Human Settlement, Philippines respectively (CSB, 1985).

France is one of the prominent silk processors of European Union countries. For centuries, the French city of Lyon has been producing fabrics of the highest quality for domestic consumption and for export as well. France is popular for designing fabrics, scarves and the fashion-houses at prominent centres like Paris. While, import share of France for silk fabrics is around 15.2 per cent and woven silk fabrics is about 14.7 per cent among the total of European Union countries. The French weaving industry which was banking on domestic supplies of raw silk is now dependent entirely on import. It imported 424 tons of raw silk in 1995 from China and the rest from Brazil, while in 1996 it amounted to 365 tons. The total import of silk and silk products has found a steady increase from 1638 tons in 1991 to 2067 tons in 1995. France has got a reputation for high quality fashionable exclusive goods. The unit prices of France export of silk goods were considerably higher as compared to Korea and Italy in 1996. The association of French silk traders (Inter - soie France) established in 1995 was actively engaged in solving the problems of international silk trade. In addition, the project EUROCHRYSALIDE gave fillip to develop high technology sericulture and to start or expand silk production in certain countries.

Historical account of sericulture shows that silk was first discovered in China and later spread to other parts of the world. Earliest record of silk was found in 2200 B.C., in public ceremony as a symbol of homage to the Emperor, Chou-King of China. The secret of silk guarded by the Chinese for about 3000 years. Korea was the second country to learn the secret of sericulture where they started the business in about 1200 B.C. According to the another version, Semiramus, a General in the army of Empress Singu-Kongo invarded and conquered Korea. Among his prisoners some were sericulturists and with the help of them he developed sericulture during the third century in Japan. During the later part of 19th century, by introducing modern machinery and intensive research techniques, Japan gave serious attention to the development of sericulture. Chinese princess carried silkworm eggs and seeds of mulberry to Tibet. Tibetians learnt the technique (Krishnswami *et al.*, 1973) and from Tibet the industry spread slowly to India and Persia, by about 140 B.C. In ancient Sanskrit literature it finds that wild silks were cultivated in India from time immemorial.

Sericulture is an ancient industry in India. References to 'silk' may be traced back to age old sacred Hindu epics. Historical records indicate the existence of a flourishing export trade in silk between India and Europe in the second century B.C. During the regime of Kanishka in 58 B.C., silk was exported from India to Rome. The industry was in flourishing stage during Mughal regime.

According to some Indian Scholars, silkworms were first domesticated in the foothills of Himalaya. The British East India Company gave the fillip to sericulture in India by developing many silk centres in different parts. Company exported large quantity of silk to England from West Bengal. Mean time, other major silk producing states like Mysore, Jammu and Kashmir developed sericulture (FAO, 1987).

Though sericulture was introduced to India in 130 B.C., it remained confined only to the Bramhaputra Vally in West Bengal and Kashmir. Later, it was Tippu Sultan who had introduced silk in Mysore, since then Karnataka has emerged as the largest silk producer in India.

Major contribution to the development of sericulture industry and silk production in India is from five states namely Karnataka, Andhra Pradesh, Tamil Nadu, West Bengal and Jammu and Kashmir. The Karnataka state is macro contributor, 64 per cent and Andhra Pradesh 18.5 per cent (Anonymous, 1998). The actual development of sericulture in Karnataka has taken five decades to attain the present status. The major advantage to Karnataka state is the farmers get timely technical support from Central Silk Board and various research Institutes located in the state.

In Karnataka, sericulture industry is 200 years old. Tipu Sultan organised the industry in 1780. Karnataka has well established technological units related to mulberry cultivation, silkworm rearing, reeling, twisting and weaving. However, industry has undergone severe variance due to pests and diseases from time to time. But, it is managed through research and extension activities. In Karnataka, out of the cultivable area of about 14.0 million ha mulberry is being grown in nearly 0.15 million ha accounting for 1.07 per cent of the cultivable area. The CSB in April 1961, took over the state research institute and in 1965 merged the training institute. The research conducted between 1961-80 has resulted in improvement and their use for hybridisation with local varieties and for bivoltine hybrids. Races like Hosa mysore, Pure mysore, NB_4D_2 were evolved. M5 mulberry variety evolved by state sericulture institute was further tested for rearing. In 1979-80 two bivoltine races namely NB_7 and NB_{18} were released by CSR & TI Mysore to replace KA and NB_7. Due to introduction of high yielding mulberry varieties, new bivoltine races, chawki rearing centres and silk production per unit area increased and renditta decreased. Silk production per ha increased from 11.98 kg to 24.35 kg and the renditta decreased from 17.5 to 13.43 kg.

The Karnataka state enjoys favourable agroclimatic conditions for expansion of sericulture activities. The activities were confined to five traditionl districts, Mysore, Kolar, Tumkur, Bangalore and Mandya. Due to confined region, the state with less spread and extension support was able to create success stories. In Karnataka, silk production per ha was 24.35 kg and silk quality was below D grade in 1980 as compared to 133.47 kg of A grade in Japan. Karnataka state was lacking certain basic factors to launch programme of production of bivoltine silk in a big way. Karnataka state launched two mega projects called KSP-I (1980-88) and KSP-II (1989-96) with world bank assistance, for better advancement of sericultural activities. However, their main objectives were to increase raw silk production by about 1600 mt of which 1000 mt was to be bivoltine. In both the projects bivoltine incentive schemes were adopted. However, the production of bivoltine silk did not catch-up as contemplated in the project. The cocoon and silk production in Karnataka during 1993-94 was 70,208 and 8,250 ton respectively. As per Das Gupta (1994), Karnataka expect 1,80,000 ha of mulberry cultivation, 1,18,630 metric tons of cocoon production and 13,732 mt of silk production by 2000-2001.

In Andhra Pradesh, nearly 80 per cent of the silk production comes from the districts of Anantpur, Chittor, Karnool and Kadappa. At present, 1.38 lakh acres of area is under mulberry cultivation and the raw silk production in the state is 2228 mt. covering 60 per cent of the farmers from small and marginal category. It is estimated that about 7 lakh families are dependent on sericulture in the state. Andhra Pradesh expects 2,00,000 acres of mulberry, 20 crores brushing of dfls, 1,00,000 tons of cocoon production including 10 per cent bivoltine, 12,500 mt raw silk production, 1500 mt bivoltine raw silk production, cocoon yield 65 kg/100 dfls, renditta for Bi × Bi 5.5 to 6, and for Multivoltine × Bivoltine 7, by the year 2000-2001 (Bhide, 1994). In Andhra Pradesh the Director of Sericulture under the Ministry of Textiles is the implementing agency for the overall development of sericulture. The State Government has also came with various schemes to give boost to sericultural activities like technical support, finance to sericulturists for workshed-cum-house for rural and urban weavers, modernization of looms, support to chawki rearing activity, marketing supports, etc.

Tamil Nadu is one of the traditional silk producing states. The state has increased its mulberry silk production over the past five decades. During 1994-95, mulberry plantation was in 13,481 ha area (Anonymous, 1998). The consumption of dfls was 107 lakh, cocoon production 7,351 mt, production of raw silk 774 mt, no. of handlooms 60,000 and requirement of raw silk was 1600 mt during 1996-97. The state government of Tamil Nadu has strong extension and promotion wing which promotes the activity and provides technical assistance. The Co-optex, a co-operative society established by reelers and weavers also plays very important role in development of silk industry by marketing raw silk and finishing silk fabric.

The sericulture activities in the state are concentrated in five districts namely, Dharampur, Madurai, Salem, Kanchipuram and Dindigad. The major districts producing cloth or *sarees* are Kanchipuram, Arani, Salem and Khumbhakenan. Presently, Tamil Nadu ranks second in India in sericultural activities (Anonymous, 1998).

In Madhya Pradesh, sericulture activity comes under the Ministry of Rural Development. The Government of Madhya Pradesh is implementing a big project on sericulture for five years with a Grant-in-Aid from Japan Economic Cooperation Fund and Technical Assistance from Japan. By this project mulberry cultivation would increase by 6000 ha and tasar activities by 2000 ha Government has also proposed to purchase land for development of mulberry and tasar sericulture. Development of market infrastructure, creation of technical service centres, one time subsidy to farmers, setting of small co-operatives, involvement of NGOs, rearing of bivoltine by adopting modern technologies from Japan and Central Silk Board are important activities of the state.

In Assam, all the four commercial varieties of silk are produced, considered to be the primary epicentre for origin and spread of sericulture in the world (Anonymous, 1988). In the total annual raw silk production of India, Assam has about 450 to 500 mt share and is one of the major silk producing state of India. Sericulture activities are carried out in almost all districts of Assam but, commercial silk weaving is concentrated in Sualkuchi of Kamrup district. This business has its roots since ages and is a traditional activity. In Assam, silk production is low due to poor technology, primitive cultural practices, non-commercialisation of cocoon production and under-developed marketing facilities. Among the four commercial varieties of silk reared in Assam, Muga is the pride of the state.

A recent survey indicates that about 20,280 villages are engaged in sericultural activity in the North-Eastern region of India. Sericulture forms a part of tradition here, especially in the states of Assam, Manipur and Tripura. Attempts have been made to develop sericulture in the hilly areas of Nagaland, Mizoram and Arunachal Pradesh, have yielded encouraging results.

West Bengal is one of the leading silk producing state in the country with a raw silk production of about 1,137 mt/year (Vijayan *et al.*, 1998). The major silk producing districts of West Bengal are : Malda, Murshidabad and Birbhum. In West Bengal, silkworms are reared during 4-5 times in a year. The climatic conditions of the state are found not conducive for bivoltine silk worm rearing through out the year. The Central Sericultural Research and Training Institute, Berhampore has identified certain genotypes for mulberry and to boost bivoltine sericulture in West Bengal.

Sericulture Industry in Kashmir is in existence since time immemorial and prior to 1897. It is traditional activity of J. & K. The World Bank Mission visited the state in March, 1988 and has sanctioned project of Rs. 40 crore. In J. & K. mainly univoltine races are practised. Presently, area under mulberry plantation is more than 2750 ha Silk reeling in Kashmir is carried out only by the government filatures. As like Kashmir in the western ghats of Maharashtra, bivoltine activities should be encouraged on large scale.

The hills of Uttar Pradesh and the sub-Himalayan tracts (Tarai region) offer ideal climatic condition for mulberry cultivation and rearing of high quality bivoltine cocoons. The annual raw silk production was 15-20 mt for the past many years in the state (Thomas, 1998). The government has proposed major silk project to produce 750 mt of high quality raw silk with raising 8500 ha of mulberry plantations.

In addition to the above-mentioned states non-traditional states like, Maharashtra, Orissa, Bihar, Gujarat, Rajasthan, Meghalaya are also taking keen interest in the development of sericulture.

In India, sericulture has already made good stride. Research contributions of CSR & TI, Mysore by way of introduction of improved mulberry varieties, development of silkworm rearing management under tropical conditions as well as introduction of productive silkworm breeds has helped to spread sericulture in wide agro-climatic areas (Datta, 1995). The production pace is slow as compared to the spread of sericulture and many sericultural states are not able to utilize the opportunity made available to them.

Inspite of significant progress in global sericulture, Indian sericulture remains primitive and is not able to come up with the production of high quality silk or fabrics in large quantity (Datta, 1995). Multivoltine breeds are popular with farmers in India (Fig. 45a and 46). The CSR & TI was successful in evolving high yielding bivoltine breeds through conventional breeding programmes. The popularity of bivoltine breeds (Fig. 43 and 45b) remains dismal due to poor survivability in the adverse tropical agro-climatic conditions coupled with diseases caused due to continuous silkworm rearing. This situation is further worsened by poor cultivation practices, rearing management, scanty ground water and soil types (Datta, 1995).

Swaminathan (1994) and Datta, (1995) says that, new opportunities for crop, tree and animal improvement by genetic combination are now available which in turn, has created new prospects for skilled employment in villages through manufacture of biofertilizer, bio-pesticides, bio-control measures and establishment of biomass refineries. As immunological melthods, hybridoma technology, plant tissue and protoplast solved culture have many problems which could not be solved through conventional approach. Hence, DBT and CSB is giving more emphasis on biotechnological research in sericulture. Several research projects have been encouraged by above agencies by providing funds. At CSR & TI, Mysore the research projects have been installed on the important aspects of sericulture namely, Restriction fragment length polymorphism (RFLP), Development of transgenic animals, analysis of genetic basis of defence mechanism in silkworm, development of immunodiagnostic kit, semisynthetic diet for young and late age silkworms, mulberry improvement through biotechnology, biofertilizers—*Azotobacter* and *Mycorrhiza* and sericultural pest management. Conventional and biotechnological approaches will help India to increase quality silk production in a big way, which will meet the increasing demand for both national and international markets. However, successful application of biotechnology in the field of sericulture science could usher a new era of high silk production in tropics in near future (Datta, 1995).

In CSR & TI, Mysore attempts have been made for semisynthetic diet for young and late age silkworms and mulberry improvement through biotechnology under which tissue culture technique and mulberry screening of genotypes for adverse situation, quick multiplications of elite genotypes, anther culture and protoplast culture and fusion, etc.

Due to the effort of CSR & TI scientists it is possible to curtail 50 per cent of recommended dose of chemical nitrogen fertilizer in mulberry cultivation through application of 20 kg of *Azotobacter* biofertilizer without adversely affecting the leaf yield and quality (Das *et al.*, 1992, 1993). It is also proved that by inoculating mulberry roots with vesicular arbuscular mycorrhizal fungi (VAM) to curtail the application of costly phosphatic fertilizer up to 75 per cent of recommended dose of phosphatic fertilizer making mulberry cultivation much cheaper (Datta, 1995). Now a days, in sericultural practices different trends are existing in India viz., low cost rearing houses (Plate II, Fig. 9), low cost shoot feeding technique (Fig. 10), shoot feeding rearing by nylon shelved, steel rack (Fig. 13), traditional wooden tray rearing (Fig. 14), etc.

As regards to mulberry protection, CSR & TI, Mysore have made tremendous progress since pests like mealy bug, *Maconellicoccus hirsutus* Green and Bihar hairy caterpillar, *Spilosoma obliqua* (Walk.) have been controlled successfully by biological means. Many natural enemies belonging to Hymenoptera, Coleoptera, Neuroptera, Lepidoptera and Diptera are reported to attack the mealy bug. The predatory beetle, *Cryptolaemus montrouzieri* Muslant feed on all stages of mulberry mealy bug. About 250 adults are required to release per acre for controlling mealy bug on mulberry garden. Technology for large scale production of this beetle has been developed at CSR & TI, Mysore. Production cost of each beetle is 27 paise (Mani *et al.*, 1987).

In general, over 300 species of insect pests have been reported on mulberry from various parts of the world (Reddi and Kotikal, 1988). From Karnataka more than 100 insect pests have been reported (Manjunath *et al.*, 1992). Recently, from Maharashtra more than 20 insect pests have been reported (Sathe and Mulla, 1999; Sathe *et al.*, 1999). In tropics loss of mulberry due to pest attack is estimated to be about 10 per cent (Sengupta *et al.*, 1990). Mealy bug, *M. hirsutus* (Hemiptera, Pseudococcidae) known to be associated with mulberry for about 75 years in eastern India, has been recorded to inflict heavy damage in southern sericultural belt in recent years (Manjunath *et al.*, 1992).

Maharashtra is considered as a non-traditional state of sericulture. Sericulture did not gain importance in Maharashtra due to alternate cropping pattern like sugarcane, cotton and other agricultural crops. However, due to international demand of sericultural products Maharashtra is also taking keen interest in this business. Sericulture in Maharashtra is not new. It is about three decades old, however, is still in its infancy.

In Maharashtra, very few sections of the society are well acquainted with Tasar and Mulberry silk production. Sericulture activities being practised on a limited scale. The state is yet to initiate organised efforts for exploring the possibility of Eri and Muga sericulture.

It is observed that several attempts have been made to introduce sericulture in Maharashtra during the 19th century. As per earlier records Government officials of the rank of collector who initiated sericulture trials, in 1827, ended with failure in 1847. Later, efforts were made but without success. It was Dr. G.P. Dewadikar an eminent botanist and agricultural scientist and cytogeneticist who made a comprehensive attempts in around 1956-57. This was possible only due to some backup or support from various corners. Mr. I.A. Kamte, initiated this efforts at the time on his own to introduce sericulture in Maharashtra and achieved good success. Then few trials on a very small scale were undertaken which proved to be very successful and encouraging. A broad based comprehensive programme then encompassing various components was drawn out.

The initial trials related to cultivation of various mulberry and non-mulberry food plants were initiated at Panchgani, Mahabaleshwar and at the foot hills of Wai. However, the industry did not spread beyond borders of Wai till 1977-78 (Tayade, 1991). Later, this was followed by cultivating the selected mulberry varieties in different agroclimatic regions in Maharashtra. Various varieties of mulberry and non-mulberry plants were collected and after finding its suitability, cultivation on a mass scale was undertaken. The attempts were also made to the comparative trials of various species of mulberry and non-mulberry to assess their relative performance under local conditions and their acclimatisation through selective breeding. Extension programmes were also drawn out based on these results which included formulation of a pattern which will be helpful to the various agriculturist. Thus, achievements were made with a view to maintain a supply of superior cuttings of mulberry to the growers and rearers.

The initiative and favourable response by a few enterprising farmers, who had opportunities to be familiar with sericulture operations of the research station have introduced sericulture activity in the state on a modest scale. The State Government took further steps to promote sericulture as it is providing employment on a large scale. (Anonymous, 1998).

In the seventh five year plan the Government of Maharashtra with the assistance of Central Silk Board (CSB) have made sincere efforts in promoting activity through Khadi and Village Industries Board as result, sericulture scattered all over the state in pockets.

The Khadi and Village Industries Board introduced few schemes under centrally assisted Western Ghats Development Plan (WGDP) and established Pilot Extension Training Centres (PETC), basic seed farm and grainage unit

(Fig. 16) in this region under which disease free layings production is done by pairing (Fig. 44) *B. mori* moths. According to Tayade, (1991) in 1990-91, 24 districts were engaged in sericultural practices and succeeded in establishing 5190 acres of mulberry plantation undertaken by 5879 farmers with production of 150 ton cocoons. While in 1997, the sericulture programme was taken up at the instance of Government of Maharashtra for mulberry cultivation in 1,862 acres by 2000 farmers of 500 villages covering 75 talukas from 23 districts in all four regions of Maharashtra (Anonymous, 1998).

During, 1998-99, the sericulture programme was taken in 2665 acres by 2635 farmers of 975 villages from 24 districts in all four agroclimatic zones of Maharashtra. However, it should not be a matter of surprise, if the mulberry cultivation area could reach up to 10,000 acres by the year 1999-2000 in Maharashtra.

Sericulture in Maharashtra has a tremendous potential, specially Western Maharashtra has conducive environment for sericulture and secondly it adjoins Karnataka, sericulturally most advanced state of India. In Vidharbha non-mulberry sericulture is well established and producing quality tasar cocoons (Fig. 47).

In recent years, biotechnological and other research work have been carried out with respect to various strains of silkworms for better cocoon production. However, the expected yield of the crop is not achieved so far because of mortalities in the worms due to natural enemies. There are several natural enemies both from insect and non-insect group to silkworms particularly, *B. mori* and *A. mylitta* (Singh and Das, 1996; Sathe, 1998). The insect natural enemies play a very important role in cocoon production as they cause higher mortalities in the worms. The mulberry silkworms *B. mori* are attacked by uzi flies, *Exorista bombycis* Louis, *Exorista sorbillans* Wiedmann, *Xanthopimpla pedator*

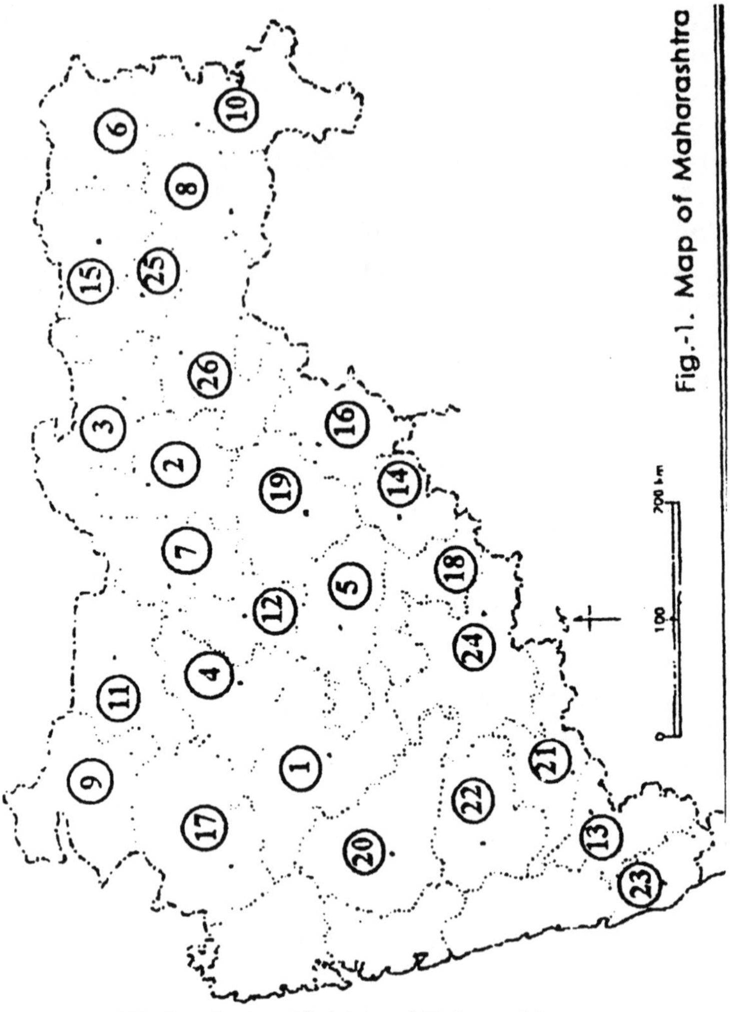

Fig.-1. Map of Maharashtra

Silk Producing Districts of Maharashtra

1. Ahemadnagar	10. Gadchiroli	19. Parbhani
2. Akola	11. Jalgaon	20. Pune
3. Amravati	12. Jalana	21. Sangli
4. Aurangabad	13. Kolhapur	22. Satara
5. Beed	14. Latur	23. Sindhudurg
6. Bhandara	15. Nagpur	24. Solapur
7. Buldhana	16. Nanded	25. Wardha
8. Chandrapur	17. Nasik	26. Yawatmal
9. Dhule	18. Osmanabad	

(Fabricius), *Apanteles glomeratus* Linn., Dermestid beetles, Praying mantids, etc. There are atleast four species of uzi flies viz., Japanese uzi fly, *Crossocosmia sericariae* (Rondani), the hime uzi fly, *Ctenophorocera pavida* (Meigen), the tasar uzi fly, *Blepharipa zebina* Walker and the Indian uzi fly, *E. bombycis* associated with silkworms, *B. mori* and *A. mylitta* and are dangerous natural enemies (Tanaka, 1964; Jolly, 1967).

The Indian uzi fly, *E. bombycis* is a primary parasitoid of the mulberry silkworm and distributed in British India (Vender Wulp, 1986), Assam, Bengal, Dehradun, Sri Lanka, Oriental region, Ethiopian and Palearctic regions (Beeson and Chatterjee, 1935), Japan (Ayuzawa *et al.*, 1972), Thailand (Ayuthaya, 1972) and Eastern Asia including West Bengal, Bihar and Assam in India extending up to Myanmar (Sriharan *et al.*, 1971). According to Krishnaswami *et al.*, (1973), the fly is very common in India, China, Thailand, South Korea and Myanmar. This fly is also distributed in Canary Islands, Oriental region, South Palearctic and Afrotropical regions, New Guinea and Queensland.

In Karnataka, the uzi fly was noticed for the first time in Bailanara Sapura village of Hoskote taluka, Bangalore district during May, 1980 (Anonymous, 1980). Earlier to this, it was confined to North-Eastern sericultural regions of the country, especially Bengal and Assam (Louis, 1980; Jameson, 1922). It is now well established in the five traditional silk producing districts viz., Bangalore, Kolar, Tumkur, Mysore and Mandya and has also spread to new areas of sericulture of the state (Rajshekargouda and Devaiah, 1983). The Indian uzi fly is also well established in some districts like Sindhudurg, Kolhapur, Sangli and Satara of Maharashtra.

In Eastern region of India, the uzi fly is known to cause 40 per cent cocoon crop losses in some seasons (Anonymous, 1982). In Malda district of West Bengal during

1962-63, over 80 per cent of cocoon production was lost due to uzi-infestation (Krishnaswami *et al.*, 1964). In traditional states like Karnataka, Andhra Pradesh, Tamil Nadu the intensity and overlapping nature of the silkworm rearings, combined with the intensive movement of cocoon over long distances and negligence of the rearers initially to destroy the affected crops helped the spread of the pest throughout the states. Several farmers lost their crop completely and continuously. Eighty per cent of the farmers' rearings were reported to be affected by Indian uzi fly in certain parts of India. The cocoon yield per 100 dfls in Karnataka was reduced to as low as 5 to 10 kg (Anonymous, 1982). Recently, in Maharashtra 100 per cent losses have been recorded in the fields of farmers and farms of State Government (unpublished data).

The Indian uzi fly *E. bombycis* is responsible for causing considerable mortalities in the silkworms, *B. mori, A. mylitta, S. ricini,* and *A. assamensis* (Cotes, 1889a, 1896; Rath, 1966; Anonymous, 1889; Thompson, 1944, Sarkar, 1980; Bharali, 1968). Thompson recorded 44 species of lepidopterous caterpillars belonging to 36 genera as hosts for *E. bombycis*. Beeson and Chatterjee (1935) listed eight species including *B. mori* as hosts of *E. bombycis*. In Japan 15 hosts have been recorded *E. bombycis* (Anonymous, 1975) which includes *A. pernyi, A. yamamai, Attacus cynthia* and *B. mori*. (Kumar and Jolly, 1986a). The taxonomic position of uzi fly is given below :

Phylum	–	Arthropoda
Class	–	Insecta
Order	–	Diptera
Family	–	Tachinidae
Genus	–	*Exorista*
Species	–	*bombycis*

As like uzi fly (*E. bombycis*), *A. glomeratus* is also found parasitizing *B. mori* larvae. Though its incidence is negligible, being a smaller sized and polyphagus in nature may be troublesome to rearers in future. *A. glomeratus* attacked many species of lepidopterous larvae (Thompson, 1953). In Japan, *Pieris rapae crucivora* Boisduval and *P. melete* Menetries (Nakata, 1963; Watanabe, 1964) are among known hosts.

The taxonomic position of *A. glomeratus* is as follows :

Phylum	–	Arthropoda
Class	–	Insecta
Order	–	Hymenoptera
Family	–	Braconidae
Sub-family	–	Microgastrinae
Tribe	–	Cotesinii
Genus	–	*Cotesia (Apanteles)*
Species	–	*glomeratus*

This parasitoid is described by Linneous and reported very first on the Pierid, *Pieris brassicae* L. in 1602 (Coppel and Martins, 1977). It is well known parasitoid of *P. brassicae* in Europe and America (Thompson, 1953). It is also reported on the same host from India in 1919 (Thompson, 1954) and from Maharashtra by Sathe (1992).

The female is characterised by having black head, thorax and abdomen; occiput nearly flat with a faintly marked median carina, front view oval; antennae slightly shorter than the body, dark reddish brown; scape thicker apically than at base; propodium regulose with a median longitudinal carina and a number of carinae (lateral) emerging from the median carina throughout its length. Legs reddish brown, hind coxa dark reddish brown, first tergite with the lateral margins parallel and only very slightly covering at apex, rugulose anterioly and striate posteriorly; second tergite slightly shorter than the third tergite, ovipositor sheath short.

A. angaleti is important larval parasitoid of Tasar silkworm in Maharashtra. It is reported from West Bengal causing 40 per cent mortalities in *A. mylitta* larvae. This parasitoid is also recorded on certain lepidopterous larvae (Thompson, 1944 and Kole and Chatterjee, 1995).

The taxonomic position of *A. angaleti* is as follows :

Phylum	–	Arthropoda
Class	–	Insecta
Order	–	Hymenoptera
Family	–	Braconidae
Sub-family	–	Microgastrinae
Tribe	–	Apantelinii
Genus	–	*Apanteles*
Species	–	*angaleti*.

The females of *A. angaleti* are characterised by having head, thorax and abdomen black, body length 2.5-3.5 mm, wing expanse 5-6 mm and ovipositor short.

Considering the importance of *E. bombycis, A. glomeratus* and *A. angaleti* as a natural enemies of *B. mori* and *A. mylitta,* they have been selected for their detail studies. The present aspects will be helpful for designing appropriate control measures of these pest parasitoids, ecological studies and making the host parasitoid index of the region.

Chapter 2
MORICULTURE

Cultivation of mulberry is called as moriculture. Mulberry leaves are used for rearing of silkworms hence they have tremendous importance in sericulture. Mulberry can be cultivated on rainfed and irrigated land. Many varieties of mulberry are in use in rearing of silkworms. Local, Kanva-2, (Fig. 3) S-36 are cultivated in irrigated conditions and local, Kanva-2, S-13, S-34 are under rainfed conditions. For bivoltine races of silkworm K2, S-54, (Fig. 4 & 8) S-36, MR-2 and S-1 varieties are found useful.

Planting (Fig. 2, 4, 6 & 7)

Mulberry can be planted variously by taking into account of variety, quality of land and climate. For bivoltine, 60 × 60 cm spacing is useful. 3 × 3 ft spacing is useful for rainfed varieties of mulberry. Planting can be done by using 6-7 inch cuttings with 3-4 healthy buds.

Fertilizer

N:P:K @ 224:250:150 kg/ha/year in 6-8 split doses should be applied. For bivoltine purpose @ 60:25:25 kg first dose/ha and 25:25:25 kg/ha 2nd dose is also applicable. However, general dose of fertilizer is 300:148:148 kg/ha/year, applied in 5 split doses.

Manuring

30-40 MT farmyard manure per ha/year in two split doses may be applied.

Harvesting of Leaves

Leaves may be harvested at morning and evening hours so that they will not dry immediately. Leaves can be harvested after 5-6 months of plantation of mulberry. In the first year 6-7 thousand kg leaves can be harvested/ acre. However, during 2nd year and onwards, after prunning, leaves can be harvested after 45 days.

Leaf Transportation

Leaves can be collected from the field with the help of basket, made up of wooden stripes. This will help in maintaining proper humidity for storage leaves.

Prunning

Prunning is essential for better quality of leaves. It is done after one year in new plantation garden and in other case, it is done before monsoon and in September-October. Prunning is done 10 inch above ground level. There are several methods of prunning which can be adopted by taking into account of age of the plant, variety and quality of land, etc. (Fig. 7).

Preservation of Mulberry Leaves

Mulberry leaves should be preserved in disinfected leaf chamber by covering wet gunny cloth in humid condition in a separate, clean and disinfected room; prevent withering and wetting mulberry. It is essential to disinfect gunny cloth with 2 per cent formalin once in 3-4 days.

Pests and Diseases of Mulberry

Mulberry is attacked by about 300 insects pests. Some more important insect pests are: scales, Jassids, mealy bugs, thrips, bihar hairy caterpiller, cutworm, stem borers, grass-hoppers, and termites etc.

Jassids, mealy bugs, scales, and thrips are cell sap suckers. By sucking the cell sap these pests affect the nutritive value of the leaves. Jassids cause hopper burnt disease, while mealy bugs cause "tukra" disease. The thrips attack tender leaves by sucking the cell sap. Jassids are 3-4 mm long, wedge shaped and greenish with dark strip on their back. The scales are circular or semi-circular black or reddish bodied found attached to the stem and leaves. Mealy bugs are wingless, only nymphs and females cause the damage to mulberry by sucking the cell sap and forming "Tukra" disease to leaves. "Tukra" is nothing but it is thickening, hardening and curling of leaves. Grasshopper and caterpillars like bihar hairy, cutworm, etc. feed on tender leaves and buds. Cerambycid beetles bore the stem of mulberry. Sucking insects and caterpillars can be controlled by foliar application of 0.2% DDVP. For mealy bugs use its natural enemy, *Cryptoleamus montrouzieri*.

Diseases

Important mulberry diseases are powder mildew, leaf spot and root knot. Powder mildew can be identified by white powdery patches on lower surface of leaves. While, leafspot by irregular brown spots on leaves. Root-knot disease is caused by nematode. The nematode form the gall on roots as a result growth of mulberry is stunted. Root-knot can be controlled by application of carbofuran 3g@30kg/ha/year in 4 splits doses. Application of pongamia or *neem* cake @ 1 ton/ha/year in 4 split doses. Powder mildew can be controlled by spraying 0.2% Dinocap or Carbendazim 50 w.p. while, leafspot by spraying 0.2% carbendazim 50 w.p.

PLATE I

Fig. 2. Mulberry garden; **Fig. 3.** Mulching in garden of M-5 variety; **Fig. 4.** S-54 variety; **Fig. 5.** Local mulberry variety with fruits; **Fig. 6.** Row system of mulberry plantation; **Fig. 7.** Paired row system of mulberry plantation and **Fig. 8.** Garden of S-54 variety.

PLATE - I

PLATE II

Fig. 9. Low cost rearing house; **Fig. 10.** Low cost shoot feeding method of rearing of *B. mori*; **Fig. 11.** Mulberry tree; **Fig. 12.** Ain plant; **Fig. 13.** Shoot feeding by nylon shelved steel rack; **Fig. 14.** Traditional wooden tray rearing method and **Fig. 14a.** Plastic mountage.

PLATE - II

Chapter 3

DISINFECTANTS

Disinfectants are the sources of destruction of the pathogens of silk worms. Destruction of disease causing pathogens is disinfection. Basically, disinfectants are of two types :

(a) Physical and
(b) Chemical

Physical disinfectants are as follows :

1. Heat
2. Steam
3. Sunlight etc.

The chemical disinfectants have more importance in sericulture as physical disinfections are comparatively less effective. The following are the important chemical disinfectants:

1. Bleaching powder
2. Formalin
3. Paraformaldehyde
4. Sodium hypochlorite
5. Lime
6. Lime and bleaching powder
7. Sodium hypochlorite and formalin mixture.
8. Bleaching powder solution and formalin lime mixture
9. R.K.O. and
10. Vijetha

Before starting rearing of silkworms, rearing equipments and rearing house should be disinfected by certain disinfectants like, bleaching powder, formalin, para-formaldehyde etc. Most widely used disinfectant in sericulture is 2% formalin (range 2-5%) and 5% bleaching powder which is easily available in the market. The disinfection can protect the silkworms from several diseases. Rearing of silkworm should not be started immediately after disinfection. One should wait at least for 24 hr for the use of disinfected house/material.

Chapter 4

SEED PRODUCTION (GRAINAGE)

Grainage is activity of production of disease free eggs/ seeds. This activity is performed by scientific community/ trained officers. The grainage activity is under control of State Governments/Government of India. In grainage activity, seeds are obtained from female moths of either pure or hybrid race of silkworms.

The silkworms get affected by certain diseases like pebrine, grasserie, flacherie and muscardine etc. Pebrine is hereditary and contaminative which is caused by a protozoan parasite called *Nosema bombycis* Nagali and is difficult to control. The inheritance of this disease is through females only. It is transmitted through stages, adult to egg to larva to pupa and to adult. It can also spread amongst the worms through contact. Once the infestation is initiated to the stock then it is very difficult to control, disposal of the stock is only solution. Therefore, pebrine is eliminated at very first stage (egg stage) of the insect in grainage. At farmer or other sericulturists level, it is practically not possible to control. Hence, grainage activities has its own importance in sericulture. Grainage activity make possible to maintain the pure stocks of seeds, and to provide seeds to farmers and rearers.

Chapter 5
REARING OF SILKWORMS

Collection of Seeds

Rearing success is based on the quality of seeds. Hence, one should take the care of seeds right from the beginning (egg laying). Seeds should be demanded from the grainage centres well in advance, i.e. 15 days to one month prior to start of rearing. The seeds collected from grainage should be fresh and each egg card should contain sufficient number of eggs. The seeds those which are not stored in freeze should be demanded for rearing purpose. Proper transportation of eggs has the great importance in rearing business. The seeds should reach safely to target spot. Hence, care should be taken that —

1. Seeds should not be kept air tight and with other material in box or in bag etc. The eggs should not be spoiled by heavy material in bags, boxes etc. Use separate bag for carrying egg cards/seeds.
2. Time of transportation plays a very important role in survival of seeds. Hence, the seeds should be transported only at evening and morning and they should not be transported at noon time as their survival and hatching is adversely affected.
3. Keep the eggs at proper place during transportation. They should not be nearer to motor engine. The bags containing eggs should not be kept at crowded places.
4. Keep the seeds safe from the rain and water during transportation.

5. After collection of eggs from grainage centre, they should be kept in trays in disinfected room.
6. The seeds should be protected from the predators like ants and wall lizards etc.
7. The seeds should be placed in trays by supporting the foampads.

Rearing of Silk Worms

In India there is tremendous scope for sericulture business as four types of silk worms namely, Mulberry silkworm, Tasar silkworm, Muga silkworm and Eri silkworms are practised. India is the only country in the world which can rear above four types of the silkworms.

Mulberry silkworms are reared on Mulberry leaves, Tasar silkworms (*A. mylitta*) on Arjuna/Som/Solu leaves and Eri silkworms on Caster leaves. Mulberry and Tasar silkworms are reared from most of the states of India but, Muga silkworms are reared only in Assam while, Eri silkworm rearing is practised in states like Assam, Bihar, Orissa and West Bengal etc.

Sericulture business has five distinct phases namely:

1. Cultivation of silkworm food plants,
2. Grainage,
3. Rearing of silkworms,
4. Reeling and
5. Marketing.

Out of above five phases of sericulture, rearing is performed by farmers. Because it is easy to them. Every member of the family can share with this business.

In any biological experimental work in the laboratory, it is necessary to maintain the culture of the experimental organism. Due to the technological advancement, modifications/alterations can be possible to improve the rearing methods of insects. Sometimes very minor changes in techniques can have drastic results, both positive and negative. Application of wrong rearing technology results 100 per cent losses in the crops. Sericulture is technology oriented job hence, rearing techniques has immense value.

A. Rearing Materials for *B. mori*

1. Rearing stand (Fig. 17)

Wooden rearing stand of the size 2.25 m in height, 1.5 m in length and 0.65 m in width consisting ten tiers are useful for keeping rearing trays.

2. Antwells

Antwells consists of enamelled plate of the 20 cm wide and 4 to 5 cm in depth and are useful for preventing ants from rearing bed (Fig. 17a).

3. Rearing tray

Rearing trays are made up of wooden material of the size 3′ long, 2′ width and 6.5 cm depth (Fig. 20).

4. Paraffin paper

It is a thick craft paper coated with paraffin wax. It is used for keeping the silkworm beds covered during rearing so as to prevent withering leaves and to maintain humidity.

5. Foam rubber strips (Fig. 36a)

Foam rubber strips of the size, 2.5 cm × 2.5 cm thick, dipped in water may be kept all around the silkworm bed in chawki rearing for maintain humidity.

6. Chopsticks

Chopsticks made of bamboo, about 17.5 cm to 22 cm long, thin in girth and tapering at the other end are used for picking delicate worms and to prevent damage by ensuring hygenic handling.

7. Feathers

Feather is used for brushing newly hatched worms and changing beds in the chawki rearing of silkworms.

8. Chopping board and knife

The chopping board of soft wood of size, 0.90 m × 0.90 m and thickness of 7.5 cm is useful for chopping the leaves in early stage with blade.size 0.3 to 0.5 m long and 3 to 7 cm broad.

9. Leaf chamber

Leaf chamber be intended for preservation of leaves in fresh condition. Leaf chamber made up of a framework of wooden strips 6.5 cm wide, spaced 6.5 cm on the sides and bottom, 120 cm × 60 cm high, all sides covered with gunny cloth is kept wet to prevent the withering of mulberry leaves stored inside the chamber for rearing.

10. Cleaning nets

The cotton nets be used for changing left over portion of mulberry leaves and litter of silkworms. Nets of different mesh size to suit various stages of silkworm are used during rearing.

11. Mountages **(Fig. 14a and 21)**

Mountage composed of a rectangular bamboo mat (Fig. 21)

on which is fixed a spiral bamboo tape of the size 5′ × 3′ with 4 to 5 cm broad tape and the space between the spirals is 4 to 5 cm. Mountage is the contrivance useful to matured worms to spin cocoons.

12. Other appliances

A hygrometer of 'Zeal' company is useful to ascertain temperature and humidity in rearing house. A room heater is useful to heat up the room for maintaining proper temperature. Disinfection pads, sprayer and wash basin (Fig. 18) are used for disinfection and washing hands with two per cent formaldehyde solution. Leaf baskets of bamboo are useful for transport of harvested leaves.

B. Rearing of *B. mori*

The silkworm eggs (Fig. 34) Pm × NB_4D_2 be collected from Government Silkworm Seed Production Centre. Incubation is done at laboratory conditions (25 ± 1°C, R.H. 75 ± 3, 18 : 6 hr., L : D). Egg hatching period is 9-10 days. Newly hatched larvae be fed (Fig. 35) with tender mulberry leaves of M5 variety and brushed with feather into wooden rearing trays of the size 3′ × 2′ (Fig. 36). At the bottom and for covering of tray, paraffin papers are used for avoiding loss of humidity during chawki stages of rearing. After and within the stage, spacing and cleaning is essential for healthy growth of silkworms. There are five instars in the larval form. The 1st instar lasts for a period of 4 to 4.5 days, 2nd instar (Fig. 37) for 3 to 3.5 days, 3rd instar (Fig. 38) 4 to 4.5 days, 4th instars 4-5 days and 5th instar 7 to 8 days. Total period required for larval development is about 27 days. Full grown larvae be kept for spinning by using bamboo mountages (Fig..21). Spinning lasts for 3-4 days. Harvested cocoons be kept in gunny bags and marketed.

PLATE III

Fig. 15 (a) Chawki rearing stand, (b) Feeding stand; **Fig. 16.** Plastic cellule for preparation of *B. mori* layings; **Fig. 17.** Rearing stand with wooden trays; **Fig. 18.** Plastic basin; **Fig. 19.** (a) Glass cage type I, (b) Glass cage type II, (c) Glass troughs; **Fig. 20.** Wooden rearing tray and **Fig. 21.** Bomboo mountage

PLATE - III

PLATE IV

Fig. 22. Rearing of *A. mylitta* by using nylon net for protection on *Terminalia* plant; **Fig. 23.** *A. mylitta* outdoor chawki rearing; **Fig. 24.** *A. mylitta* fifth instar larva; **Fig. 25.** *A. mylitta* indoor rearing by using bottles; **Fig. 26.** *A. mylitta* third instar rearing and **Fig. 27.** *Spodoptera litura* larva

PLATE - IV

Rearing of silkworm is divided into two phases :

1. *Chawki rearing* : or rearing of young age silkworms. It is rearing of silkworms from 1st to 2nd instars.
2. *Late age silkworm rearing* : It includes, rearing of worms from 3rd instar to last instar.

Brushing

Brushing is collection of newly hatched worms from egg cards to the convenient place. Brushing is done by feather so that worms would not die due to injury. Uniform hatching is essential for easy rearing and better results. Hence, at blue stage of eggs they are kept in black box for uniform hatching. Black box stops the development of blue eggs.

For rearing of silk worms environmental conditions, spacing, feeding time and dose, instars and their duration, leaf size and quantity, bed size and cleaning etc. are given in the Table 4b.

Care at Moulting

During moulting period, the worms appear with a very typical position, like a snake with erected head. During moulting the worms stop feeding. They take about 15-30 hr for completion of moulting in general. If irregular moulting is observed then the worms may be grouped accordingly for further rearing. The beds should be kept dry for avoiding pathogenic attack. The worms may be dusted with RKO/ceresan lime (for newly hatched).

Mounting and Harvesting

At the end of 5th age instar, worms stop feeding and seek out place for spinning the cocoon. If proper place is not available, the worms will start spinning on unwanted places. Hence, mountages should be provided at proper

Table 4b: Requirements for Rearing 100 dfls

Instar of worm	Duration (days)	Feeding time per day	Feeding time Total	Feeding dose Total (kg)	Leaf proportion (gms.)	Leaf number on plants	Leaf size	No. of trays (3'×2'×4½") Initial	End	Tray cleaning time	Duration of shedding cuticle (hr)	Humidity (%)	Temp. (°C)	Bed size (Bi/multi × Bi) hybrids
First	3-3.5	3-4	11	2-2.5	200-250	From top 3-6	0.5 to 2 cm^2	2	4	2	20	85	26-28	4-15 sq.ft
Second	2-2.5	3-4	10	6-7	600-700	7-12	2 to 4 cm^2	4	8	2 Before & after moult	24	80	26-28	15-45 sq.ft
Third	3-3.5	4-5	14	25-30	2-2.5 kg	11-15	4-6 cm^2	8	16	3, After, during & before moult	24	80	25	46-90 sq.ft
Fourth	4-4.5	4-5	19	75-85	4-4.5 kg	Medium/ June	Entire leaf	16	32	Once every morning	30	70-75	24-25	91-180 sq.ft
Fifth	6-7	4-5	31	600-700	19-20 kg	Medium/ June	Entire leaf	32	64	Once every morning	—	70	23-24	181-360 sq.ft

PLATE V

Fig. 34. Egg of *B. mori.* (Pm × NB_4D_2); **Fig. 35.** Newly hatched larvae of *B. mori.*; **Fig. 36.** Chawki rearing of *B. mori.*; **Fig. 36(a).** Foam rubber strip; **Fig. 37.** Second instar larvae of *B. mori.*; **Fig. 38.** Third instar larvae of *B. mori*; **Fig. 39.** Cocoon of *B. mori.* (Pm × NB_4D_2); **Fig. 40(a).** Male pupa of *B. mori.*; **Fig. 40(b).** Female pupa of *B. mori.*; **Fig. 41(a).** Female adults of *B. mori.* and **Fig. 41(b).** Male adults of *B. mori.*

PLATE - V

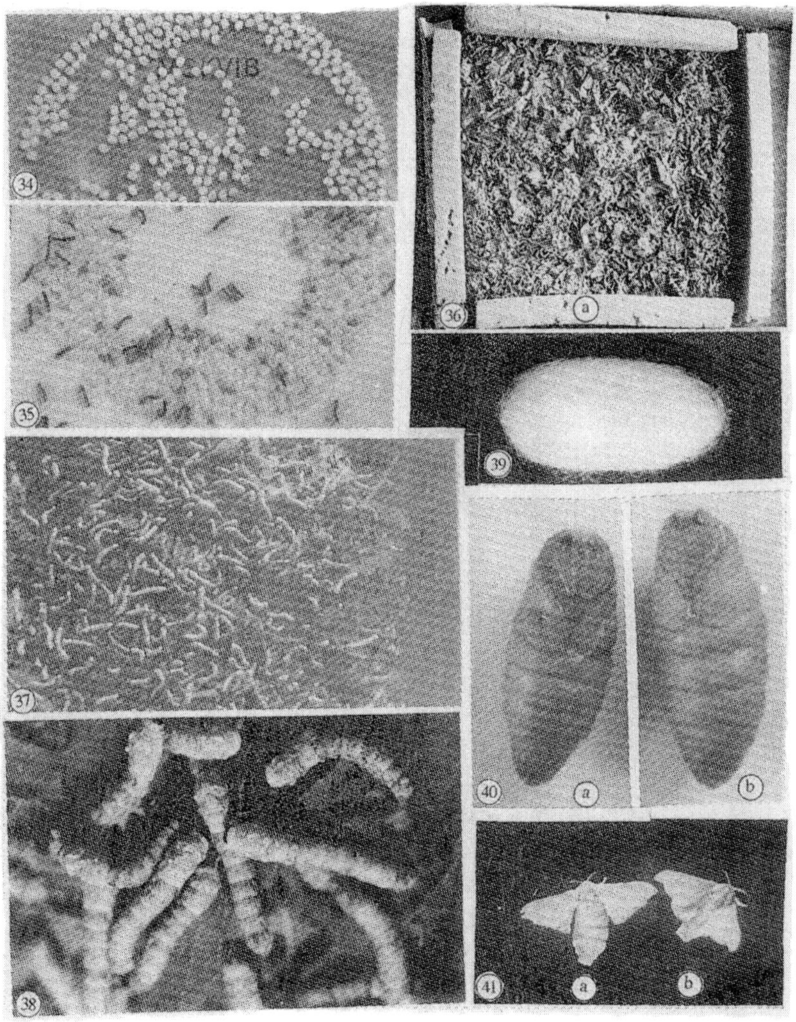

PLATE VI

Fig. 42. (Pm × NB_4D_2) multivoltine cocoon; **Fig. 43.** Bivoltine cocoon; **Fig. 44.** Mating of *B. mori* adults; **Fig. 45(a).** Multivoltine cocoon of *B. mori.*; **Fig. 45(b).** Bivoltine cocoon of *B. mori.*; **Fig. 46.** Multivoltine *B. mori* cocoon and **Fig. 47.** Local tasar cocoons.

PLATE - VI

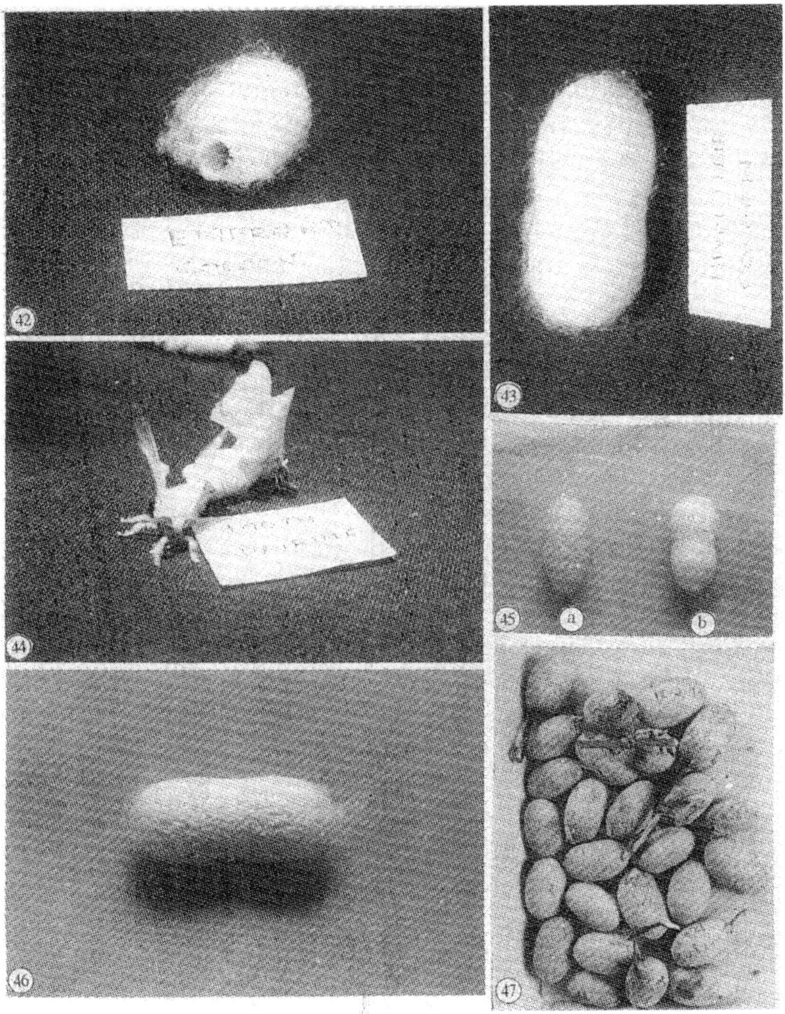

time. The ripen worms are collected and put on the montages/"chandrike". During cocooning stage, cocoons should be protected from the urination of worms, cocoons get stained, and should be provided proper atmospheric conditions.

The cocoons can be harvested on the 5th or 6th day of spinning in general. Too many days delay in harvesting will result pierced cocoon formation. As soft cuticle of pupae raptures due to bad handling, harvesting should not be done immediately after pupation. The pupal skin should be hard at the time of harvestation.

Transportation of Cocoons

Transportation of cocoons should be done with great care that the bags/gunny bags should not filled tightly. Do not mix good quality cocoons with other. Protect cocoons from heavy weight and water during transportation. Give proper aeration to cocoons during transportation.

Several hybrids are in practice in India, ex. Pm × NB_{18}, NB_{18} × P_5, My_1 × NB_{18}, P_2D_1 × NB_{18}, SH_6 × KA, CA_2 × NB_4D_2, RD_1 × NB_{18}, SH_6 × NB_4D_2, etc. Certain varieties are region specific and season specific. NB_{18} × P_5 is practised in Orissa and M.P. in summer, CA_2 × NB_4D_2 in West Bengal, Orissa and M.P. in spring; NB_{18} × P_5 in Orissa and M.P. in winter and RD_1 × NB_{18} in U.P. in summer/early autumn.

Pests and Diseases

Silkworms are attacked by several pests, the list of which is given in the text. The most important pests of *B. mori* are uzi-flies and dermestid beetles. Uzi-flies are similar to house-fly but they are much bigger than the house-fly. Three species of uzi-flies are recorded on *B. mori*, they are listed below :

Indian uzi-fly	—	*Exorista bombycis*
		E. sorbillans and
Japanese uzi-fly	—	*Sturmia sericarae.*

Indian uzi-fly is blackish gray with four longitudinal black bands on thorax. The females are 10.2 mm long while, the males 11.9 mm long. The fly deposit her eggs on the silkworm body resulting black scars. After hatching from the eggs, the maggots start feeding internal tissues of the silkworm, moult three times inside the host body and finally the third instar break the body wall/cocoon wall (if infestation is from 4th to 5th instars) and kill the silkworm/pupa. Cocoon threads are broken into several pieces thus, becomes useless for reeling.

The uzi-flies can be controlled by adopting nylon or wire mesh to windows or doors or ventilators. The uzi flies can also be controlled by using uzi tab and uzi trap. Biological control of uzi flies is possible by using their parasitoid, *Nesolynx thymus,* one lakh females and 500 males in three splits release @ 8000, 16000 and 16000.

Dermestids are small sized beetles, about 1 cm. long oval elongated, dark brown with clubbed antennae. These beetles feed on cocoons at storage. They can be controlled by fumigation of methyl bromide and chloropicrin, or by passing hot air, 50-60 °C in the stored house.

Diseases

Pebrine, grasserie, flacherie and muscardines are the important diseases of *B. mori*. Pebrine is eliminated at grainage. Hence, now a days it is not troublesome. But, grasserie and flacherie are quite troublesome.

Grasserie

Grasserie is caused by *Borrelina* virus. Infected worm

becomes yellowish and intersegmental membranes become swollen, particularly middle segments of the body. Skin becomes fragile, ruptured easily and release milky blood.

Control

Use proper leaves and provide proper ventilation. Avoid extreme low and high temperature. RKO can be sprinkled on the worms before brushing and after each moult.

Flacherie

It is caused by bacteria, *Streptococci, Coli* and *Bacillus*. Infected larva becomes sluggish, do not feed, in 5th instar diarrhorea manifests, body becomes black and putrefies. The disease can be controlled by feeding appropriate quality and quantity of leaves. Disinfect the rearing house and equipments. Disinfect the bed with lime and bleaching powder mixture.

Pebrine

The larva appears with black spots on body, becomes sluggish and dull, larvae develop unequally. Larvae moves slowly, lack luster and turn rusty brown. The pebrine can be controlled at grainage level, however 2 per cent formalin is effective for all purpose of disinfection.

C. Rearing of Tasar Silkworm :

Rearing Material for *A. mylitta*

1. Nylon net

Nylon net of the mesh size 20 mm be used to protect larvae from predators by applying over 3 feet height pruned food plants (Fig. 22).

2. Rearing set

Rearing set consist of bottles, wooden trays, polythene paper etc.

3. Rearing bottles

Rearing bottles of the size 20 cm height, 9 cm width and 2 cm mouth are used to keep twigs of plants for rearing of the early stage tasar silkworms (Fig. 25).

4. Ain twigs

About one and half feet long Ain twigs consisting of tender leaves are used for feeding early stage silkworms (Fig. 26).

5. Polythene papers

Polythene papers are used in wooden trays consisting bottles along with Ain twigs for maintaining hygenic conditions and also to cover the twigs so as to prevent from early withering.

6. Cotton

Cotton is used to plug the mouth of bottle to protect the larvae from drowning and also to check an increase in humidity due to gradual evaporation of the water.

Rearing of *A. mylitta*

The eggs (Fig. 48) are collected from Regional Tasar Research Station, Bhandara/Laboratory. The eggs are allowed to hatch and the 1st instar larvae (Fig. 49) are reared in the laboratory on the leaves of Ain trees (*Terminalia tomentosa*) subsequently, 3rd, 4th and 5th instars (Fig. 24) are released on Ain trees (Fig. 22) in the field. Larvae be transferred from one tree to other when foliage is exhausted.

The observations are made on the formation of cocoons (Fig. 53) and further adult (Fig. 54) emergence.

The life cycle of *A. mylitta* is completed in a period of 58 to 60 and 65 days for first and second crop season, respectively. During both the crop seasons, the incubation period averaged eight days. The larval development requires 35 and 40 days while the pupal period extends over 15 to 18 days and 18 to 20 days for the first and second crop, respectively.

Chawki/Control Rearing Method for *A. mylitta*

Until the first moult the larvae are reared indoor by using water filled bottle (Fig. 25) holding 3 to 5 tender quality foliage twigs (about 40-50 cm long) (Fig. 25). The cut end of the twigs be inserted well under water. This arrangement keep the foliage turgid for 2 to 3 days for feeding. The mouth of the bottle be plugged by cotton to protect the larvae from drowning and also to check an increase in humidity due to gradual evaporation of the water. The rearing capacity can be increased considerably by using a big plastic bucket or container (Fig. 23). The larvae coming out of moult are allowed to crawl onto fresh twigs and transferred outdoors (Fig. 22). Other details are similar as described earlier for these worms. Sometimes 3rd instars are reared indoor (Fig. 26).

The life cycle of *A. mylitta*

Stage	First crop July to Sept. Period in days	Second crop Sept. to Oct. Period in days
Egg	8	8
Larva -I Instar	3	3
II Instar	4	4
III Instar	6	7
IV Instar	6	7
V Instar	9	11
Pupa	17-20	17-20
Adult longevity (control)		
Male	3-4	4-5
Female	5-6	5-6

PLATE VII

Fig. 48. Eggs of *A. mylitta*; **Fig. 49.** Newly hatched larva of *A. mylitta*; **Fig. 50.** Second instar larva of *A. mylitta*; **Fig. 51.** Fifth instar larva of *A. mylitta*; **Fig. 52.** Pupa of *A. mylitta* in cocoon; **Fig. 53.** Cocoon of tasar and **Fig. 54.** Adult of *A. mylitta*

PLATE - VII

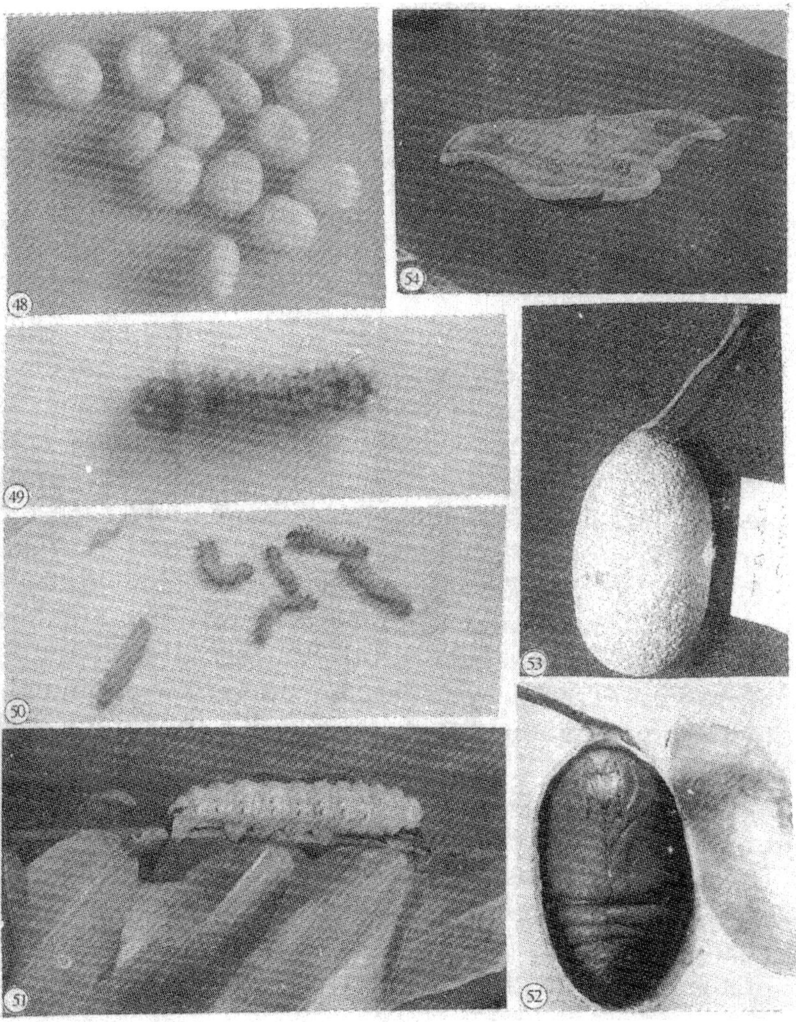

Chapter 6

SURVEY OF NATURAL ENEMIES OF SILKWORMS

Introduction

Sampling insect population is one of the important task in population dynamics which estimate the number of the species present in the target area. For sampling the insects with specialised habits like parasitoids and predators requires basic knowledge on biological peculiarities. In general, the parasitoids are usually specific for one or closely related group of insects. Thus, they are restricted to, and dependent on an individual of host species. Their reproductive rates are also related to the availability of their hosts. Thus, parasitoid sampling is the process of sampling of the pest species. The parasitoids frequently determine the population densities of the pest species. In general, the survey and surveillance of pest and their parasitoids has tremendous economic importance in the pest management. Survey studies or estimates of number of parasitoids per unit of land area are therefore necessary to construct life tables and for nearly all other studies related to population dynamics of the species in the field conditions.

The tachinid fly, *Exorista bombycis* (Louis), popularly known as uzi-fly is a well known endoparasitoid of mulberry silkworm *Bombyx mori* Linn. in the western region of Maharashtra particularly in the districts of Kolhapur, Sangli, Satara and Sindhudurg. The parasitoid probably might be introduced from Karnataka, very 1st at Chandgad and Gadhinglaj regions of Kolhapur district during the year 1990-1991 (unpublished data). However, the incidence of parasitoid is not recorded officially on *B. mori* (Pm × NB_4D_2) up to date. Recently, the regional joint director sericulture of Maharashtra intimated about the incidence and its control. In Maharashtra *B. mori* is also attacked by a small braconoid wasp, *Apanteles glomeratus* Linn. (Hymenoptera : Braconidae) with low per cent parasitism but, uzi-fly menace is increasing day by day considerably. The uzi-fly cause damage to mulberry silkworms and cocoons.

In Maharashtra, tasar silkworm rearing has also great importance as it is practised in Vidarbha region. The activities for long time were governed by Development Corporation of Vidarbha Ltd. (Nagpur) and Maharashtra State Khadi and Village Industries Board. Recently, the Maharashtra Government under the Ministry of Textile has formed separate directorate for overall development of sericulture in the state. However, the expected out come is not achieved so far in sericulture because of the various types of pests and natural enemies and diseases of silkworms. The mulberry and tasar silkworms in Maharashtra are attacked by several natural enemies, the list of which is given in Table 5 and 6. In Maharashtra the tasar silkworms are parasitized by the hymenopterous braconoid wasp, *Apanteles angaleti* Muesebeck, in addition to the uzi-fly, *Bhlepharipa zebina* Walker (Sathe, 1998). *E. bombycis*, *A. glomeratus* and *A. angaleti* are important insect natural enemies of silkworms in Maharashtra, hence they have been selected for their survey and surveillance studies. The review of literature indicates that Thompson (1944, 53, 54), Chatterji and Swarup (1961), Krishnaswami *et al.* (1964), Datta *et al.* (1977), Venkatadri (1980), Anonymous (1980, 1981, 1982, 1989), Hu (1982), Jolly and Chandrashekharaih (1982); Rajashekhargouda and Devaiah

(1983), Sukhatme *et al.* (1984), Choudhury (1984), Murty *et al.* (1988), Gangavar and Thangavelu (1989), Barman *et al.* (1990), Govindraju *et al.* (1990), Siddappaji and Basavanna (1990), Mahadevappa (1992). Narsimha *et al.*, (1993, 1995), Kole and Chatterjee (1995), Chakraborty *et al.*, (1996a and 1996b), Menezes (1997) etc. attempted survey studies on natural enemies of lepidopterous worms.

The present study will be helpful for understanding the distribution and ecology of the silkworms and also making the host parasitoid index of the region. Keeping in view all above facts, the present topic was selected.

Material and Methods

Survey of natural enemies of *B. mori* were conducted at Chandgad and Gadhinglaj blocks in Kolhapur district from 1995 to 1997. The observations covers 5 rearing crops in randomly selected 10 per cent of the total commercial sericultural villages (Surute, Devarwadi, Chennekupi, and Harli). Ten dfl silkworms rearing batches from each (10) farmers were examined once during the 4th and 5th instars for uzi-fly infestation, based on the number of black scars on the silkworm larvae. The infected larvae were separately reared for adult emergence.

The survey of *A. angaleti* was conducted at Ajara Distt. Kolhapur, Maharashtra, visiting every ten days. The larvae were collected for a fixed time of an hr at Ajara forest on Ain (*Terminalia tomentosa* W. and A.) and screened for parasitoid cocoons, allowing them to feed on the Ain leaves, under controlled conditions, nylon nets on the prunned plants of Ain. The detailed rearing procedure of *A. mylitta* is described in chapter 1-5. The cocoons or adult parasitoids emerged from infected silkworms were collected and identified and preserved by pinning/keeping in 4% formalin. Other natural enemies both from vertebrates and invertebrates were also surveyed by spot observation and tabulated in Table 5 and 6.

Results

1. Survey of *E. bombycis*

The survey of *E. bombycis* was made at farmers field with ten dfls at Kolhapur (Chandgad-Devarwadi, Surute and Gadhinglaj-Chennekupi, Harli) blocks. The average data of survey study indicated that the uzi incidence was started from January in regular rearing stock and to the seasonal crop from first week of September. In general, the incidence of the parasitoids peaked in June and it declined to September and again increased in October. The incidence of uzi again found increased in the month of February and gradually decreased in summer. The results are tabulated in Tables 7, 8 and 9 and Figs. 29, 30 and 31.

2. Survey of *A. glomeratus*

The survey studies of *A. glomeratus* showed that its incidence was noticed in the month of April and May, except these two months incidence was not observed on the crop throughout the year. The observations on the incidence are recorded in Tables 7, 8 and 9 and Figs. 29, 30 and 31.

3. Survey of *A. angaleti*

The results on survey of *A. angaleti* are tabulated in Tables 10, 11 and 12 and in Figs. 32 and 33. The results indicate that the population of *A. angaleti* peaked in the month of October, incidence was started from the month of August, however, it was found decline after October and from December onwards the incidence was not observed.

Table 5 : Natural Enemies of Mulberry Silkworm *B. mori*

Sr. No.	Common Name	Generic Name	Family	Order	Class	Occurrence
(a)	**Predators**					
1.	Dermestid beetles	*Dermestis cadverinus*	Dermestidae	Coleoptera	Insecta	Sept. to Dec.
2.	Praying mantid	*Hierodula bipapilla*	Mantidae	Dictyoptera	Insecta	July to March
3.	Earwings		Forficulodae	Dermaptera	Insecta	June to Nov.
4.	Wasps	*Polistes hebraeus*	Vespidae	Hymenoptera	Insecta	June to Nov.
5.	Ants	*Oecophylla smaragdina*	Formicidae	Hymenoptera	Insecta	June to Nov.
6.	Sting bug	*Canthecona furcellata*	Pentatomidae	Hemiptera	Insecta	June to Nov.
7.	Ruduviid bug	*Sycanus collaris*	Reduviidae	Hemiptera	Insecta	
(b)	**Parasitoids**					
8.	Ichneumonid	*Xanthopimpla pedator*	Ichneumonidae	Hymenoptera	Insecta	Sep. - Feb.
9.	Braconid fly	*Apanteles (Cotesia)/ glomeratus*	Braconidae	Hymenoptera	Insecta	Sep. - Feb.
10.	Braconid fly	*A. belippae*	Braconidae	Hymenoptera	Insecta	Sep. - Feb.
11.	Braconid fly	*Á. stantoni*	Braconidae	Hymenoptera	Insecta	Sep. - Feb.
12.	Braconid fly	*A. plutellae*	Braconidae	Hymenoptera	Insecta	Sep. - Feb.
13.	Uzi fly	*Exorista bombycis*	Tachinidae	Diptera	Insecta	Thr. out Year
(c)	**Vertebrates**					
14.	Cobra snake	*Naja naja*	Elapidae	Squamata	Reptilia	Nov - June
15.	Rat snake	*Ptyas mucosus*	Elapidae	Squamata	Reptilia	Nov - June
16.	Wall lizard	*Hemidactylus flaviviridis*	Gekkonidae	Squamata	Reptilia	Thr. out Year
17.	Garden lizard	*Calotes versicolor*	Agamidae	Squamata	Reptilia	Thr. out Year

(a)	**Aves**					
18.	Owl	*Bubo nipalensis*	Strigidae	Strigiformes	Aves	Thr. out Year
19.	House sparrow	*Passer domesticus*	Ploceidae	Passeriformes	Aves	Thr. out Year
20.	Jungle crow	*Corvus splendens*	Corvidae	Passeriformes	Aves	Thr. out Year
21.	Bulbul	*Phycnonotus jocosus*	Pycnonotidae	Passeriformes	Aves	Thr. out Year
22.	Indian Myna	*Acridotheres tristis*	Stumidae	Passeriformes	Aves	Thr. out Year
23.	Wagtail	*Motacilla alba*	Motacillidae	Passeriformes	Aves	Thr. out Year
(b)	**Mammalia**					
24.	Jackal	*Canis aureus*	Canidae	Camivora	Mammalia	Thr. out Year
25.	Fox	*Pteropus giguanteus*	Canidae	Camivora	Mammalia	Thr. out Year
26.	Monkey	*Macoca mulatta*	Ceropitheadae	Primate	Mammalia	Thr. out Year
27.	Mouse	*Mus musculus*	Muridae	Rodentia	Mammalia	Thr. out Year
28.	Rat	*Rattus rattus*	Muridae	Rodentia	Mammalia	Thr. out Year
29.	Squirrel	*Funambulus palmarum*	Sciuridae	Rodentia	Mammalia	Thr. out Year

Table 6 : Natural Enemies of Tasar Silkworm *A. mylitta*

Sr. No.	Common Name	Generic Name	Family	Order	Class	Occurrence
(A) Insecta						
(a) Predators						
1.	Pentatomid bug	*Canthecona furcellata*	Pentatomidae	Hemiptera	Insecta	July to Oct.
2.	Raduviid bug	*Sycanus collaris*	Reduviidae	Hemiptera	Insecta	August to Oct.
3.	Praying mantid	*Hierodula bipapilla*	Mantidae	Dictyoptera	Insecta	July to March
4.	Wasps	*Vespa* sp.	Vespidae	Hymenoptera	Insecta	July to March
5.	Red ant	*Oecophylla smargdina*	Formicidae	Hymenoptera	Insecta	July to March
(b) Parasitoid						
6.	Ichneumonid fly	*Xanthopimpla pedator*	Ichneumonidae	Hymenoptera	Insecta	Nov., Jan.
7.	Braconid fly	*Apanteles angaleti*	Braconidae	Hymenoptera	Insecta	Aug., Nov.
8.	Uzi fly	*Blepharipa zebina*	Tachinidae	Diptera	Insecta	July to Oct.
(B) Vertebrates						
(a) Reptiles						
9.	Rat snake	*Ptyas mucosus*	Elapidae	Squamata	Reptilia	July to Jan.
10.	Garden lizard	*Calotes versicolor*	Gekkonidae	Squamata	Reptilia	Thr. out Year

(b) Aves						
11.	Bulbul	*Pycnonotus jocosus*	Pycnonotidae	Passeriformes	Aves	Thr. out Year
12.	Jungle crow	*Corvus splendens*	Corvidae	Passeriformes	Aves	Thr. out Year
13.	Indian Myna	*Acridotheres tristis*	Sturnidae	Passeriformes	Aves	Thr. out Year
14.	House sparrow	*Passer domesticus*	Ploceidae	Passeriformes	Aves	Thr. out Year
(c) Mammalia						
15.	Monkey	*Mococa mulatta*	Ceropitheadae	Primate	Mammalia	Thr. out Year
16.	Mouse	*Mus. musculus*	Muridae	Rodentia	Mammalia	Thr. out Year
17.	Rat	*Rattus rattus*	Muridae	Rodentia	Mammalia	Thr. out Year
18.	Squirrel	*Funambulus palmarum*	Sciuridae	Rodentia	Mammalia	Thr. out Year

Table 7 : Survey of *E. bombycis* and *A. glomeratus* on *B. mori* at Kolhapur* 1995.

Month	No. of Larvae brushed (10 dfls)	No. of Larvae parasitized by		No. of Parasitoids emerged		Per cent parasitism	
		E. bombycis	*A. glomeratus*	*E. bombycis*	*A. glomeratus*	*E. bombycis*	*A. glomeratus*
January	3151	221	0	119	0	7.01	0
February	3398	408	0	307 (10)**	0	12.00	0
March	3058	108	0	86	0	3.53	0
April	3491	0	1	0	78	0	0.028
May	3131	200	2	120	169	6.3	0.063
June	3221	1127	0	1003 (130)	0	34.98	0
July	3678	0	0	0	0	0	0
August	3110	49	0	21	0	1.57	0
September	3010	0	0	0	0	0	0
October	3511	980	0	703(10)	0	27.91	0
November	3462	311	0	197(20)	0	8.98	0
December	3661	0	0	0	0	0	0

* Average data of four centres (Surute, Devarwadi, Chennekupi and Harali).

** The figures in parenthesis indicate number of pupal parasitoids.

Table 8 : Survey of *E. bombycis* and *A. glomeratus* on *B. mori* at Kolhapur* 1996.

Month	No. of Larvae brushed (10 dfls)	No. of Larvae parasitized by		No. of Parasitoids emerged		Per cent parasitism	
		E. bombycis	*A. glomeratus*	*E. bombycis*	*A. glomeratus*	*E. bombycis*	*A. glomeratus*
January	3350	63	0	38	0	1.88	0
February	3367	299	0	177(31)**	0	8.88	0
March	3089	0	0	0	0	0	0
April	3120	131	0	89	0	4.19	0
May	3298	98	1	58(2)	103	2.97	0.030
June	3748	862	0	625	0	22.99	0
July	3268	0	0	0	0	0	0
August	3881	0	0	0	0	0	0
September	3492	123	0	91	0	3.52	0
October	3156	347	0	187 (13)	0	10.99	0
November	3668	261	0	207 (13)	0	6.92	0
December	3424	0	0	0	0	0	0

* Average data of four centres (Surute, Devarwadi, Chennekupi and Harali).

** The figures in parenthesis indicate number of pupal parasitoids.

Table 9 : Survey of *E. bombycis* and *A. glomeratus* on *B. mori* at Kolhapur* 1997

Month	No. of Larvae brushed (10 dfls)	No. of Larvae parasitized by		No. of Parasitoids emerged		Per cent parasitism	
		E. bombycis	*A. glomeratus*	*E. bombycis*	*A. glomeratus*	*E. bombycis*	*A. glomeratus*
January	3691	0	0	0	0	0	0
February	3313	380	0	201	0	11.46	0
March	3268	116	0	91	0	3.54	0
April	3621	0	1	0	98	0	0.027
May	3312	0	3	0	218	0	0.09
June	3808	1029	0	874 (74)**	0	27.02	0
July	3214	0	0	0	0	0	0
August	3301	0	0	0	0	0	0
September	3434	213	0	139	0	6.29	0
October	3719	520	0	279 (36)	0	13.98	0
November	3374	0	0	0	0	0	0
December	3219	0	0	0	0	0	0

* Average data of four centres (Surute, Devarwadi, Chennekupi and Harali).

** The figures in parenthesis indicate number of pupal parasitoids.

Table 10 : Survey of *A. angaleti* on *A. mylitta* at Ajara 1995-96

Month	Date	No. of Larvae Collected	No. of Larvae Parasitized	No. of Parasitoids Eerged	Per cent Parasitism
July	1	1	0	0	0
	10	0	0	0	0
	20	3	0	0	0
August	9	0	0	0	0
	19	4	1	63	25.00
	29	5	1	79	20.00
September	8	0	0	0	0
	18	9	2	54	22.22
	28	10	1	92	10.00
October	8	12	1	102	8.33
	18	18	3	278	16.66
	28	4	1	109	25.00
November	7	0	0	0	0
	17	6	1	78	16.66
	27	1	0	0	0
December	7	1	0	0	0
	17	2	0	0	0
	27	0	0	0	0
January	6	1	0	0	0
	16	2	0	0	0
	26	0	0	0	0

Table 11 : Survey of *A. angaleti* on *A. mylitta* at Ajara 1996-97

Month	Date	No. of Larvae Collected	No. of Larvae Parasitized	No. of Parasitoids Emerged	Per cent parasitism
July	1	0	0	0	0
	10	0	0	0	0
	20	2	0	0	0
August	9	0	0	0	0
	19	4	0	0	0
	29	4	1	110	12.5
September	8	0	0	0	0
	18	8	1	62	12.5
	28	9	0	0	0
October	8	13	0	0	0
	18	8	2	136	25.00
	28	14	4	321	28.57
November	7	10	1	69	10.00
	17	0	0	0	0
	27	2	0	0	0
December	7	0	0	0	0
	17	0	0	0	0
	27	1	0	0	0
January	6	0	0	0	0
	16	2	0	0	0
	26	1	0	0	0

Table 12 : Survey of *A. angaleti* on *A. mylitta* at Ajara 1997

Month	Date	No. of Larvae Collected	No. of Larvae Parasitized	No. of Parasitoids Emerged	Per cent Parasitism
July	1	0	0	0	0
	10	0	0	0	0
	20	0	0	0	0
August	9	0	1	72	0
	19	6	1	0	16.66
	29	9	0	0	0
September	8	2	0	0	0
	18	7	1	97	14.28
	28	8	1	79	12.50
October	8	6	1	103	16.66
	18	13	3	312	23.7
	28	16	2	165	12.50
November	7	0	0	0	0
	17	1	0	0	0
	27	10	1	53	10.00
December	7	2	0	0	0
	17	0	0	0	0
	27	1	0	0	0

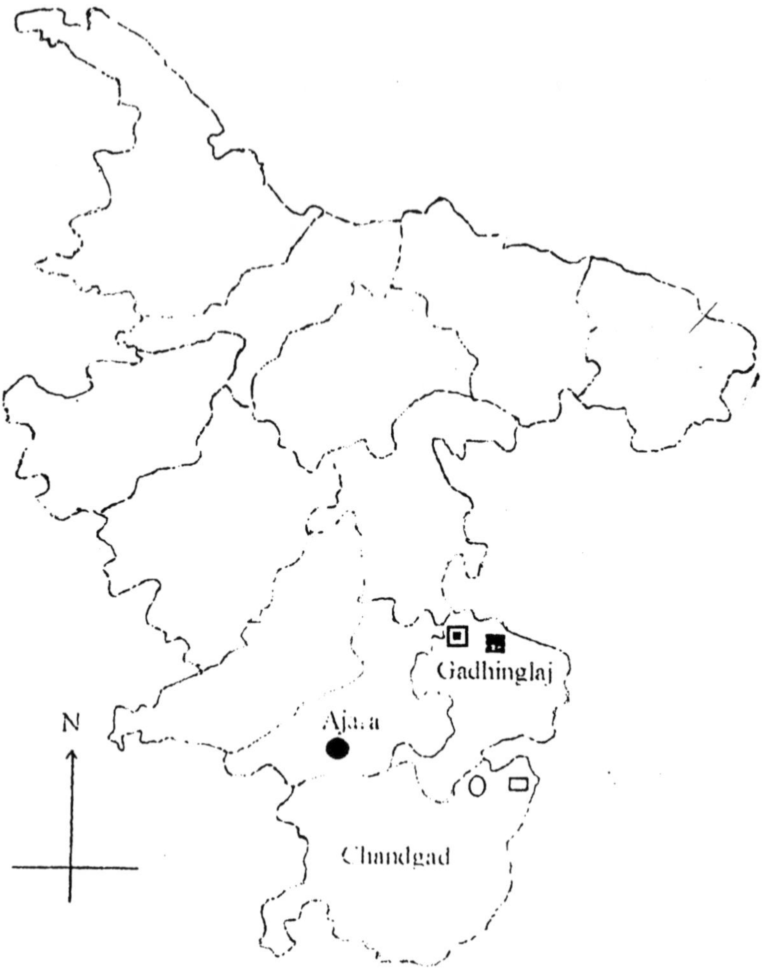

Survey Study Centres

Ajara	(a) Ajara	●
Chandgad	(a) Devarwadi	○
	(b) Surute	□
Gadhinglaj	(a) Chennekupi	▣
	(b) Harali	■

Fig. 28 : Map of Kolahpur District

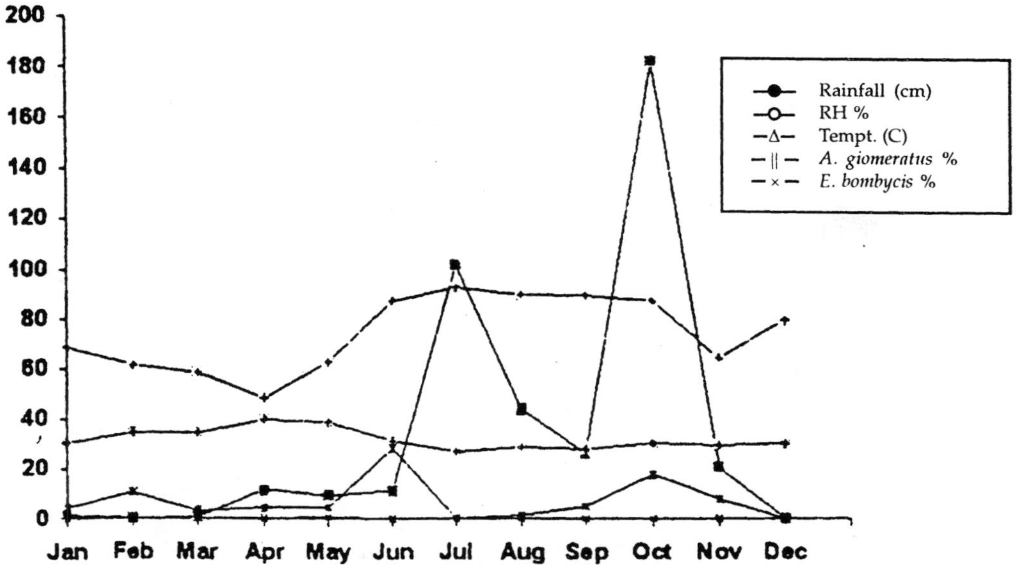

Fig. 29 : Seasonal abundance of parasitoids of *B. mori* with respect to meteorological data collected at Chandgad during survey study for the years 1995, 96 and 97.

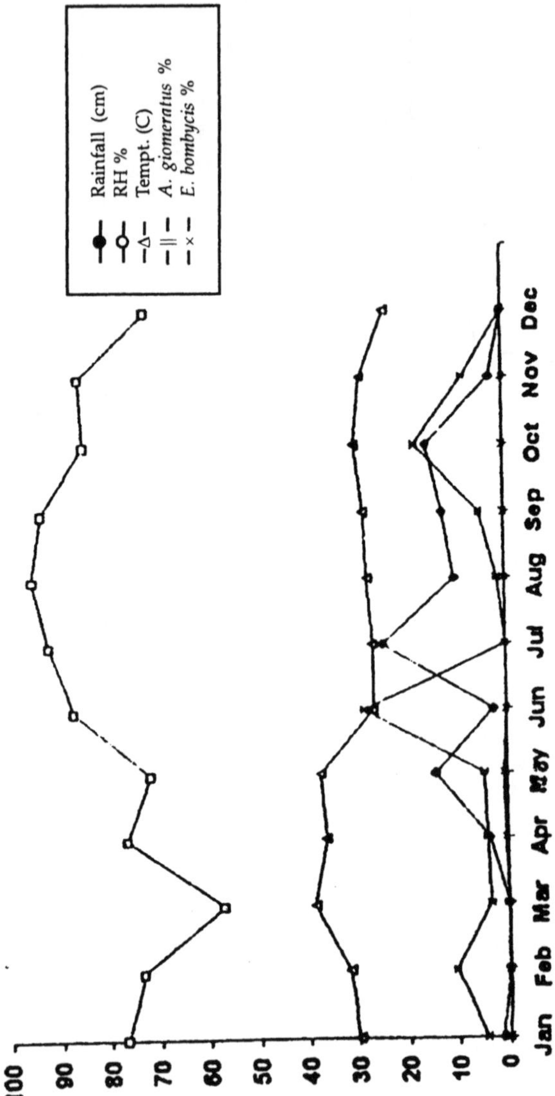

Fig. 30 : Seasonal abundance of parasitoids of *B. mori* with respect to meteorological data collected at Gadhinglaj during survey study for the years 1995, 96 and 97.

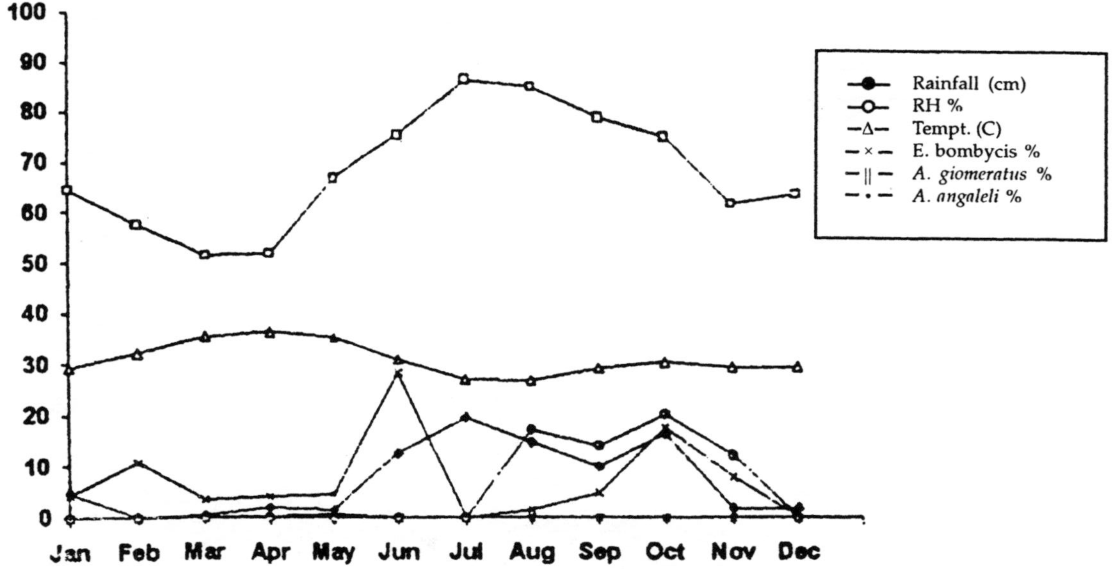

Fig. 31 : Average seasonal abundance of parasitoids of *B. mori* and *A. mylitta* at Kolhapur during the years 1995, 96 and 97.

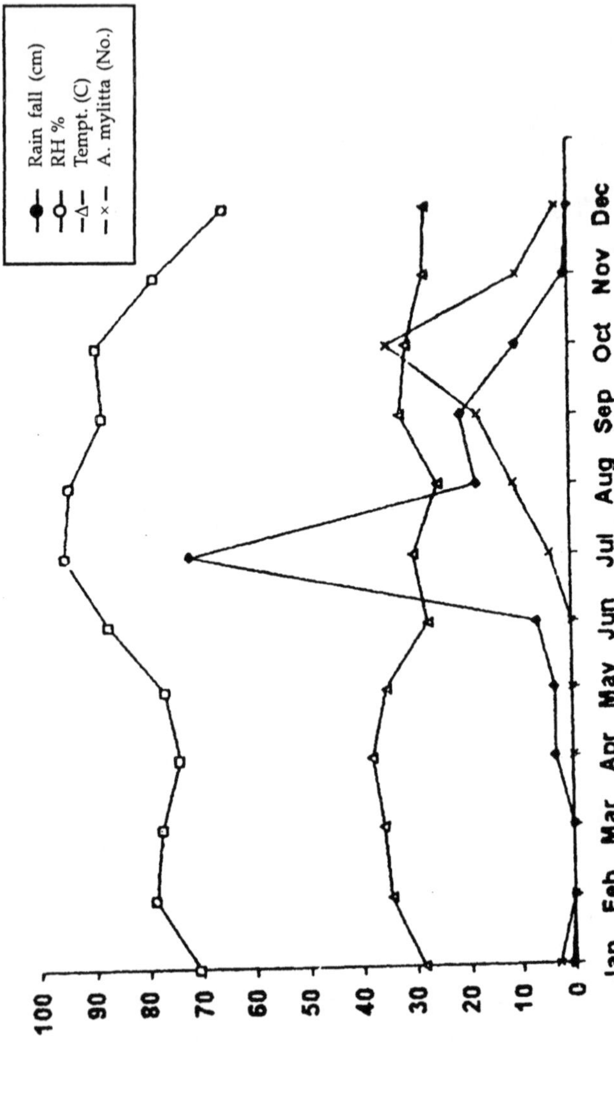

Fig. 32 : Seasonal abundance of *A. mylitta* with respect to meteorological data collected at Ajara during survey study for th years 1995, 96 and 97.

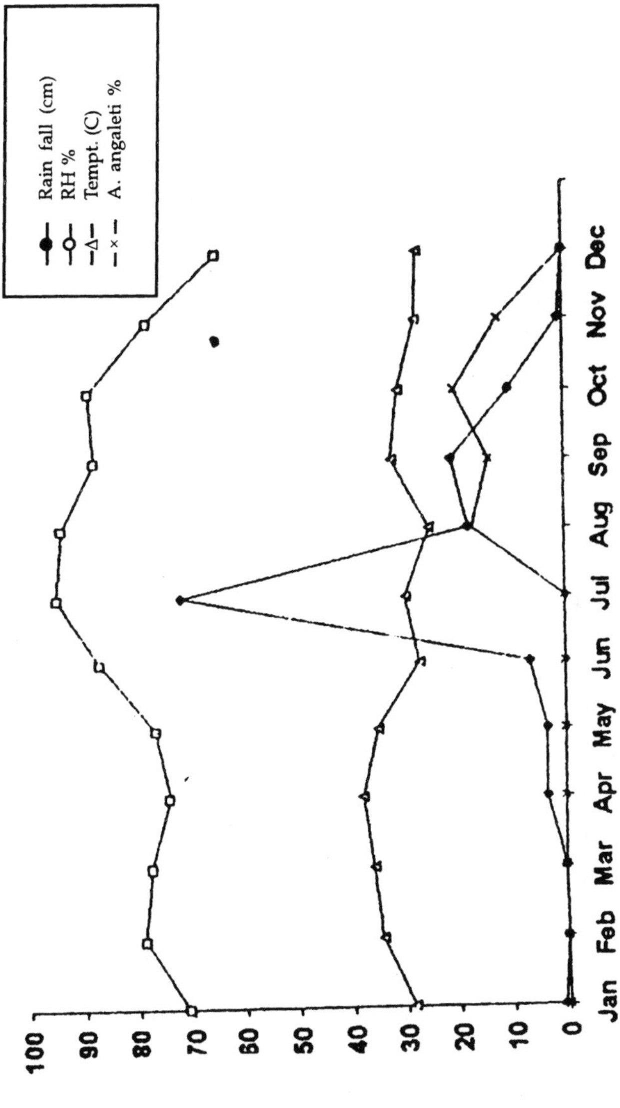

Fig. 33 : Seasonal abundance of *A. angaleti* with respect to meteorological data collected at Ajara during survey study for the years 1995, 96 and 97.

Discussion

In Maharashtra, the survey studies of *E. bombycis, A. glomeratus* and *A. angaleti* have not been carried out previously and this is first attempt. As per the experts opinion from sericulture department of Maharashtra, the probable incidence of uzi was during 1990-91 from the bordering villages like Surute and Dewarwadi of Kolhapur district and might came from Karnataka as it is well established there. Now, it is spread throughout the Western Maharashtra, particularly in the districts of Kolhapur, Satara, Sangli and Sindhudurg (unpublished data). Many a times sericulture department purchased seed-cocoons from traditional sericultural belts of Karnataka. Probably, the uzi might came along with the seed cocoons. Another important reason of incidence of uzi in Maharashtra might be the natural population in the nature as it breed on several agricultural pests like Bihar hairy caterpillar *Spilosoma obliqua* (Walk), Tobacco caterpillar *Spodoptera litura* (Fab.), Gram pod borer *Helicoverpa armigera* (Hubn.), etc.

The uzi-fly, *E. bombycis* is dangerous pest of several strains of *B. mori*. Hence, its survey studies have been made from different parts of India by many workers (Anony-mous, 1980, 1981, 1982; Jolly and Chandrashekharaiah, 1982; Choudhary, 1984; Gangawar *et al.*, 1989; Chakraborty *et al.*, 1996 a, b, etc.)

The sporadic appearance of the black scar over the body of silkworm larvae which is assuredly symptomatic to uzi-fly infestation was first noticed in some sericultural pockets of Karnataka during early 1980, in Bylanarasipura village of Bangalore district (Anonymous, 1980). The incidence was also noticed in Kutgal in Karnataka during 1981 (Anonymous, 1981).

Tricholyga bombycis Beck is a serious pest of *B. mori*, found attacking all stages of silkworms in Assam. Incidence of this fly was very high in summer (Choudhary, 1984). Reports of uzifly infestation to silkworm 1st and 2nd stages in Tamil Nadu have been exposed by Gangawar and Thangavelu (1989). Govindraju *et al.* (1990) reported that in Karnataka, uzifly infestation varied from 9.4 per cent to 19.9 per cent. The infestation reached peak in October, 1987 and September, 1988 which had highest rainfall and relative humidity in those years. Uzifly infestation reached to lower extremes during July, 1987 and February, 1988 coinciding with low rainfall and relative humidity (Govindraja *et al.*, 1990).

In southern part of Karnataka in the year 1981-82, Jolly and Chandrashekharaiah (1982) studied the case of about 26000 sericulturists and reported that about 19-22 per cent of the farmers lost 2-4 crops in a year whereas, about 6-10 per cent of the farmers lost 5-6 crops. Reports are also available from Karnataka as this fly pest was introduced through Nistari seed cocoons brought from West Bengal by unscrupulous seed producers in 1980 (Mahadevappa, 1992). Inspite of best efforts put forth by the department to completely eradicate the fly pest, due to non-cooperation from the rearers the fly population increased and spread rapidly. The extent of damage to the silkworm crop in some of the heavily infested areas has reached about 40 per cent. With the introduction of some of the preventive measures to combat the uzi menaces presently, the extent of damage due to uzi fly is reported to be in the range of 10 to 30 per cent (Mahadevappa, 1992).

Uzi-fly is particularly serious in West Bengal where the loses to the cocoon crops has been reported to be very serious. Assessment of the seasonal incidence of the pest indicated that uzi was very serious in September-November and preferred later stages than chawki (Krishnaswami *et al.*, 1964). In Assam and West Bengal *E. sorbillans* was reported as a serious larval and pupal parasitoid of *B. mori*

(Rajshekargouda and Devaiah, 1983). Uzi-fly has been also reported in southern districts of Karnataka but it was not reported from northern part (Rajshekhargouda and Davaiah, 1983).

Barman *et al.* (1990) studied the incidence and control of uzi-fly, *T. bombycis* in Bangaladesh, they found that maximum incidence of uzi was in April, May and June and further declined in August, September. The pest was controlled with the help of eucalyptus oil.

Chakraborty *et al.* (1996b) studied the seasonal incidence of uzi fly, *E. bombycis* infestation to the silkworm *B. mori* and assessment of cocoon loss in Birbhum district, West Bengal, India during the period, February, 1991 to November, 1992 and reported that the rate of infestation at the fourth and fifth larval instars and at cocoon stage was in relation to temperature, humidity and rainfall. The investigations indicated a positive significant correlation coefficient among these factors. The maximum uzi-fly infestation was 16.1 per cent in July at the fifth instar; the lowest infestation rates 0.44 per cent and 0.36 per cent were observed at the fourth instar in February and at cocoon stage in November respectively. Based on multiple regression equations and respective coefficients of determination, they have predicted prevailing abiotic conditions for future uzi-fly infestation, has been calculated and was found to be maximum of 30.82 per cent in Birbhum district.

Chakraborty *et al.* (1996a) also studied the incidence of uzi-fly infestation to *B. mori* and assessment of cocoon to crop loss in Murshidabad district of West Bengal from February, 1991 to November, 1992 under 10 rearing seasons in relation of the aboitic factors (temperature, humidity and rainfall). Their observations showed a positive significant correlation coefficient among the factors. Minimum uzi fly infestation was 0.12 per cent in February,

1991 at 4th instar and maximum infestation of 7.30 per cent was recorded in July, 1991 at 5th instar. A multiple regression equation was fitted for predicting the rate of uzi fly infestation in different seasons. Estimation of cocoon crop loss by uzi fly attack in Murshidabad District showed crop loss to an extent of 19.70 per cent.

Bhattacharya *et al.* (1993) reported maximum infestation (11.6%) of uzi fly in May during 5th instar in Malda district. Chakraborty (1996a) reported maximum infestation (7.30%) in July during 5th instar in Murshidabad district. While the present investigations showed maximum infestation (28.33%) during 5th instar in Kolhapur district in the month of June.

General trend of incidence of uzi infestation with respect to different commercial rearing seasons reflects that infestation being low, during November to February, gradually increased with rise in temperature, humidity and rainfall in following seasons. The infestation was recorded maximum during rainy season (June to August). In the studies of Chakraborty *et al.* (1996a) the extent of cocoon crop loss due to uzi infestation in Murshidabad district was higher (19.07%) than in Malda district (12.25%) while in the present study, the infestation reached in its peak in the rainy season (June to August) and observations were made on the 4th instar. In Kolhapur district some farmers are engaged in continuous rearing of the silkworms through-out the year. Thus, the uzi fly on emergence finds host of suitable stages for ovipostion. However, in Chennekupi, Dewarwadi places of Kolhapur district the farmers are maintaining the gap of 2 to 3 months between each crop (unpublished data) that reduced rate of uzi infestation in these centres.

In present study, uzi infestation reached in its peak (28.33%) in rainy season (June). Again it was found increased (17.62%) in winter season (October). The per cent parasitism decreased as compared to October and February, 10.78 per cent and it was almost negligible in summer months, particularly in April and May. These results are in agreement with the statements of Chakraborty *et al.* (1996a). The increase in percentage of uzi infestation in June may be due to change in climatic conditions like increase in temperature, humidity and rainfall. The another probable reason for increase in uzi infestation to mulberry silkworms may be due to scarcity of its natural alternative hosts as at this time most of the *kharif* agricultural crops get harvested, similarly, increase in infestation of uzi in June may be probably due to harvest of summer crops.

As regards the *A. glomeratus* no such reprots are available on the survey studies pertaining to *B. mori* infestation. This pest is reported on *B. mori* in 1894 in India by Thompson (Chatterjee and Swarup, 1961).

The incidence of *A. glomeratus* on *P. brassicae* was studied by Sood *et al.* (1995). In *P. brassicae* the mortality begins due to *A. glomeratus* from 2nd fortnight of March. The gregarious endoparasitoid, *A. glomeratus* which killed the larvae of *P. brassicae* in the 5th instar, remained associated till early May and resulted into the mortality ranging from 3.9 to 23.6 per cent with peak parasitization in the 2nd week of April. The mortality recorded by Sood *et al.* (1995) is in the range of 40 per cent on *P. brassicae*. In the present study its incidence on *B. mori* was reported in April, May. The incidence of this parasitoids in these months may be due to scarcity of its original hosts or high population and over crowding, may lead to the parasitoid to search for *B. mori*. At present its infestation is negligible (0.061%) but, may increase in future due to the availability of abundant hosts, *B. mori* and may cause the problem to sericulture industry.

As regards to the *A. angaleti* survey on *A. mylitta*, a very little is known as Kole and Chatterjee (1995) reported this parasitoid from West Bengal, infesting the tasar 2nd instar larvae upto 40 per cent during September. The mortalities recorded in worms were about 5-10 per cent in 2nd stage, 15-2 per cent in 3rd stage and 10.15 per cent in 4th and 5th stages. However, these workers have not reported seasonal abundance of this parasitoid. In the present study infestation performance to the stages of tasar worms is in agreement with the reports of above workers and the survey studies showed that their incidence was maximum (20.35%) in October on Ain (*T. tomentosa*) at Ajara of Kolhapur districts (M.S.). The present work will be helpful for ecological studies of parasitoids and silkworms, making host parasitoids index of the region and designing control measures of the parasitoids studied.

Chapter 7

ASPECTS OF SILKWORM PARASITOIDS

I. LIFE HISTORY

Introduction

Thousands of the parasitic wasps occur throughout the world, possessing complex and fascinating biologies, and frequently function as determinants of pest population densities (Doutt, 1959); because of the last attribute, the parasitoids are of tremendous economic importance and therefore widely used in biological pest control programmes (Coppel and Mertins, 1977).

The parasitoids are entomophagous insects which differ from true parasites by following features : (i) They do not exhibit heteroecism, (ii) The host is usually of the same taxonomic class, i.e. Insecta, (iii) In population dynamics their action resembles with predators more than that of true parasites, (iv) In comparison with their hosts, they are of relatively large size, (v) The development of an individual destroys its host, (vi) They are parasitic as larvae only, the adults being free living forms.

The parasitoids are widely scattered in several orders of insects like Hymenoptera, Diptera, Lepidoptera, Hemiptera etc. Among the parasitic orders, Hymenoptera ranks first since several families are parasitic in nature.

More than one lakh species have been described only from Ichneumonidae and Braconidae families of Hymenoptera. From Diptera, FamilyTachinidae is very prominent which contains more than one thousand and five hundred parasitic species.

Taxonomical assessments in parasitoids were made by the adult characteristics. The reliable understanding of the species would be more correct by studying their immature forms and life cycle. Life history aspects also correlates to probable phylogeny of the groups. The review of literature showed that Thorpe (1932), Ishikawa (1934), Fulton (1940), Short (1952), Juillet (1960), Gangrade (1964), Kajita and Drake (1969), Cardona and Oatman (1971), Datta and Mukherjee (1978), Patil and Govindan (1984), Siddappaji (1985), Sathe (1986a, 1987a, 1988a, b, 1990a, 1992), Sathe *et al.* (1988a) etc. studied the life history aspects in parasitoids from Hymenoptera and Diptera. The data on life cycle will be helpful in understanding the development, morphological peculiarities of the immature forms and authentic identity of the species. Moreover, the life history aspects has tremendous importance in mass rearing of the species needed in pest management programme.

In the present topic, *Exorista bombycis* Louis and *Apanteles glomeratus* Linn., the parasitoids of *Bombyx mori* Linn. (Pm × NB_4D_2) and *Apanteles angaleti* Muesebeck, a parasitoid of *Antheraea mylitta* Drury have been considered for the life history studies. *E. bombycis* is previously attempted by Devaiah *et al.* (1993), Datta and Mukharjee (1978) etc. with respect to life cycle. Likely, some preliminary aspects of life history of *A. angaleti* have been reported by Kole and Chatterjee (1995). While in case of *A. glomeratus* no reports are available on the life cycle with host, *B. mori*. *A. glomeratus* was studied with respect to other host, *Pieris brassicae* Linn. by Osmolousky (1964) and *E. bombycis* is considered here for its repetitive work for confirming the results.

Material and Methods

To study the life history of above parasitoids, initial culture of them and their hosts were maintained in the laboratory as mentioned in the chapter 7-II. Life cycle in all above parasitoids were studied by exposing 10- day old host larvae to them. Host density for parasitization exposure was 100 in each case, the host density were exposed to the respective parasitoids for 24 hr and sufficient number of parasitized hosts were obtained. The parasitized host larvae were dissected regularly every day up to 40 days and sufficient number of immature stages were collected for morphological observations. Simultaneously, life cycle was studied by keeping the record of egg lying period, incubation period, larval period, pupal period and adult longevity.

Results

1. Life cycle of *E. bombycis*

Life cycle consists of egg (Fig. 56), maggots : 1st instar (Fig. 58), 2nd instar (Fig. 59), 3rd instar (Fig. 60), pupa (Fig. 61) and adult (Fig. 62) and completed within 32.9 days. The detailed description of all the stages of the fly and duration was given by Ischikawaa (1993), Sanan (1934) and Devaiah *et al.* (1993). The duration of life cycle with different stages of 20 individuals of *E. bombycis* under laboratory conditions (25±1°C, R.H 75±3, 18:6 hr. L:D) are represented in Table 13.

2. Life cycle of *A. glomeratus*

The details of duration of different stages of life cycle of *A. glomeratus* are given in Table 14.

Table 13 : Developmental Period and Longevity of *E. bombycis*

Sr. No.	Egg period (days)	Larval period (days)	Pupal period (days)	Adult emerged on (days)	Adult longevity with 50% honey soln (days)	Sex
1	3	5	10	18	10	M
2	4	9	15	28	12	M
3	5	11	22	38	19	F
4	3	8	24	35	21	F
5	4	10	19	33	16	F
6	4	12	17	33	14	F
7	3	8	21	32	15	M
8	6	10	19	35	17	F
9	4	7	23	34	20	F
10	3	8	22	33	12	M
11	4	9	20	33	11	M
12	5	11	19	35	18	F
13	4	13	21	38	15	F
14	3	10	19	32	13	M
15	3	11	18	32	12	M
16	5	9	19	33	16	F
17	4	10	20	34	15	F
18	4	11	16	31	9	M
19	7	13	21	41	19	F
20	4	11	15	30	8	M
Average	4.1	9.8	19	32.9	14.6	

Egg

The newly deposited eggs were white, thin walled, stalked and typically hymenopteriform (Fig. 64).

Instars

The parasitoid has 3 instars, 1st two were vesiculate and last was hymenopteriform.

1st Instar

The 1st instar was opaque with translucent body, consisting quaderatly pointed head, three thoracic and seven abdominal segments and characterized by caudal vesicle (Fig. 65).

Table 14 : Developmental Period and Longevity of *A. glomeratus*

Sl. No.	Egg pericd (days)	Larval period (days)	Pupal period (days)	Adult emerged on (days)	Adult longevity with 50% honey soln (days)	Sex
1	2.5	5	4.5	12	6	M
2	2	5	4	11	7	M
3	2	7	5	14	10	F
4	2	7	4	13	10	F
5	2.5	7.5	4	14	11	F
6	2	6.5	4.5	13	11	F
7	2	6	5	13	9	M
8	2	7	4	13	7	M
9	2.5	7	4.5	14	11	F
10	3	6	5	14	12	F
11	2	6	4	12	6	M
12	2	7	4	13	8	M
13	2.5	6	4.5	13	9	F
14	2	6	5	13	9	F
15	2	7	5	14	11	F
16	3	6	4	13	9	F
17	3	6	4	13	7	M
18	2	7.5	4.5	14	9	M
19	2	6	5	13	8	M
20	2	6	5	13	10	F
Aver-age	2.25	9.375	4.475	13.1	9	

2nd Instar

The body of the 2nd instar was cylindrical, more or less straight and white and was with caudal vesicle. The larva consisted 13 well defined segments, spiracles were not observed.

3rd Instar

The 3rd instar larva is with typical hymenopteriform which tapered slightly towards both the ends (Fig. 66). Early third instar was with caudal vesicle but it was withdrawn in the matured forms. The larval segments were prominent in matured larvae. The larvae came out of the host body by killing them. Newly emerged parasitoid larvae spin silvery cocoon colony (Fig. 67), the cocoons were silvery and were rounded at both the ends.

Pupa

The pupa of *A. glomeratus* was exarate or free type. Initially it was light yellow except blackish eyes and brownish ocelli later, the pupa became brownish.

Adults

Adults are black in colour measuring in about 3 mm in body length and characterized by having the propodeum with areola, ovipositor very short and more or less concealed (Fig. 68).

3. Life cycle of *A. angaleti*

The life cycle of *A. angaleti* consists of egg, larva, pupa and adult. The life cycle was completed in an average period of 13.6 days. The details of duration of immature forms of 20 individuals are tabulated in Table 15.

Egg

The eggs were deposited into the host body with the help of sharp ovipositor. The newly deposited eggs were white, thin walled and rounded at both ends but, one end was slightly tapering (Fig. 73).

Table 15 : Developmental Period and Longevity of *A. angaleti*

Sr. No.	Egg period (days)	Larval period (days)	Pupal period (days)	Adult emerged on (days)	Adult longevity with 50% honey soln (days)	Sex
1	2	6	5	13	9	M
2	2.5	7.5	5	15	13	F
3	2	5	4	11	7	M
4	3	7	5	15	12	F
5	2.5	6	5.5	14	7	M
6	3	5	5	13	8	M
7	2	6.5	4.5	13	8	M
8	2	7.5	4.5	13	13	F
9	2.5	6.5	5	14	12	F
10	2.5	7.5	5	15	14	F
11	2	5.5	4.5	12	9	M
12	3	6	5	14	11	F
13	3	6	5	14	12	F
14	2	6	5	13	9	M
15	2	7	4	13	7	M
16	2	7	5	14	8	F
17	3	6	5	14	12	F
18	3	6	4	13	8	F
19	2.5	6.5	5	14	11	F
20	3	7	4	14	13	F
Average	2.475	6.375	4.95	13.55	10.15	

1st Instar

The newly hatched larvae were mandibulate, vesiculate and whitish in colour. Their body was translucent showing head quadrate and 3 thoracic and 7 abdominal segments. After two days larvae become more opaque (Fig. 75).

2nd Instar

Second instar was whitish in body coloration and cylindrical straight. The head was narrow in second instar and larvae were with 13 well defined segments and a prominent anal vesicle (Fig. 76).

3rd Instar

This instar was characterised by having 13 clear body segments and slightly tapering shape towards both the ends. The mouth parts were heavily sclerotized and well developed. In the matured form vesicle was lacking. These larvae found emerged from the host with large numbers, about 100-150. The emerged parasitoid larvae constructed a silken cocoon colony (Fig. 77).

Cocoon

Cocoons were densely spun, cylindrical and rounded at both ends. The parasitoid is gregarious (Fig. 78).

Pupa

As like *A. glomeratus,* pupa of *A. angaleti* was exarate type. It was creamy white initially but, became brownish black at the time of completion. Its eyes were blackish throughout the development.

Adults

Adult fly is blackish in colour, about 2.5 to 3.5 mm in length and 5-6 mm in wing expance. The female (Fig. 72) is with short ovipositor and male (Fig. 74) smaller than the female.

Discussion

The Indian uzi-fly, *E. bombycis* is a primary parasitoid of the mulberry silkworm *B. mori* distributed in India and is studied by several workers (Ghosh, 1933; Beeson and

Chatterjee, 1935; Thompson, 1944; Sriharan *et al.*, 1971; Ayuzawa *et al.*, 1972; Ayuthaya, 1972; Krishnaswami *et al.*, 1973; Anonymous, 1980; Louis, 1980; Anonymous, 1982; Rajashekhargouda and Devaiah, 1983; Vender Wulp, 1986 etc.). Recently, Devaiah *et al.* (1993) studied the life cycle of *E. bombycis* and reported that the fly has completed 5-8 generations in a year, one generation was completed in 15.4 days to 50 days depending on season. The hatching of egg was in 1.5 hours to 3-4 days. The larval period was ranged from 4 to 7 days. The pupal duration was ranged from 7 to 8.5 days. The adult male survived for 5 to 15 days and female for 20 to 25 days. In the present study *E. bombycis* was studied at laboratory condition (25±1°C, RH 75±3, 18:6 hr, L:D) by exposing 10-day old *B. mori* larvae on which the parasitoid completed its life cycle within 32.9 days while, the another braconoid parasitoid, *A. glomeratus* completed its life cycle in 13.1 days with the same laboratory conditions.

In *Campoletis chlorideae* Uchida, a parasitoid of *Helicoverpa armigera* (Hubner) there were five instars (Gangrade, 1964) likely, in *Diadegma trichoptilus* Cameron five instars were noted (Sathe, 1988a). But, in present forms only 3 instars were noted.

Broodryk (1969) observed 3 instars in *Chelonus* (*Microchelonus*) *curvimaculatus* Cameron. The first occupied the greatest part of the larval life and the other two instars were completed in a relatively short period. The same situation occurred in *Cotesia diurinii* R. & N. (Sathe, 1986a). Fulton (1940) reported 3 instars in *Apanteles congregatus* (Say). Similarly, in the present braconoid parasitoids three instars were observed. In *A. dignus* Muesebeck, the total developmental period from egg to adult was 18 days (Cardona and Oatman, 1971) While, in *A. creatonoti* Viereck, a larval parasitoid of *Thiocidas postica* Wlk. it was 14-15 days (Sathe *et al.*, 1988). Whereas in the present forms,

A. glomeratus and *A. angaleti* the life cycles were completed in 13.1 and 13.6 days respectively which are shorter in duration than the above two parasitoids.

Orgilus obscuratus (Nees) (Hymenoptera, Braconidae) had five instars rather than three, including three mandibulate forms (Juillet, 1960) but, in the present three parasitoids, there were only three instars. Kajita and Drake (1969) studied the biology of *Cotesia chilonis* (Munakata) (=*Apanteles chilonis*) and *Cotesia flavipes* (Comeron) (=*Apanteles flavipes*) the larval parasitoids of *Chilo supptessalis* Walker and reported three instars in them. The 1st two were vesiculate and the third one was hymenopteriform. Similarly, Cardona and Oatman (1971) reported three instars in *Apanteles dignus* (Muesebeck). The first was being mandibulate and other two were hymenopteriform. In the present braconoid forms, only third instar was hymenopteriform.

The life cycle studies of above parasitoids will be helpful for mass rearing of these parasitoids and understanding the development of the parasitoids. It will also help in adopting the control measure strategies of them.

II. REARING OF PARASITOIDS

Rearing Material

1. Glass cages

The glass cage consists of wooden base and glass wall on three sides. In one type (Fig. 19a) the 4th side was closed by muslin cloth with a slive for handling the insects and in the other type (Fig. 19b) the 4th side was with glass window. These cages were used for the studies like oviposition, host specificity and host age selection and also for host plant parasitoid interactions.

2. Glass troughs

The glass troughs used for rearing of silkworms/ parasitoids were of two types viz., 9 and 12 cm in height and 20 and 25 cm in diameter respectively. The caterpillars that were collected from the field have been screened for their parasitoids in these troughs. For the prevention of escape of parasitoids, the glass troughs were covered by muslin cloth (Fig. 19c and d).

3. Plastic containers

Plastic containers of the size 1.5 ft diameter and petridishes of two sizes 18.5 cm and 9 cm in diameter were used for handling rearing of eggs, parasitized/ unparasitized larvae and pupae of hosts.

4. Specimen tubes

For keeping the parasitoid cocoons for adult emergence, specimen tubes of 10 cm and 5 cm in height and 2 cm in diameter were used. The open ends of tubes were covered with muslin cloth for good ventilation.

Rearing of Parasitoids

1. Rearing of *E. bombycis*

The pupae (Fig. 61) of uzi flies were collected from the cocoon markets and kept in the plastic containers/specimen tubes/test tubes for individual adult (Fig. 62) emergence. Newly emerged pair (1 male : 1 female) were caged in test tube, size 19 × 2.5 cm for mating. After mating the females were exposed to the 4th instar *B. mori* larvae (Fig. 56) for parasitization in insect cage type 1 (Fig. 19a). After parasitization (Fig. 55), the parasitized larvae (Fig. 57) were separated and reared in wooden tray by usual procedure of silkworm rearing. Within 9 to 10 days the maggots of uzi start emerging from the silkworms by breaking the body wall and killing them. The 3rd instar (Fig. 60) maggots were collected and kept in glass troughs (size 9 cm height and 20 cm in diameter) (Fig. 19c) containing a thin layer of soil for pupation in about 10 to 12 hr the maggots became pupae. Later, the pupae were kept in specimen tubes or small plastic containers individually for adult emergence. During the rearing of this parasitoid the laboratory conditions were 25±1°C, 75±3 R.H. and 18:6 hr. L:D. The adult parasitoids were fed with 50% honey and the *B. mori* with M-5 mulberry leaves.

2. Rearing of *A. glomeratus*

The cocoons of *A. glomeratus* were collected form the fields of cabbage (Fig. 84) on *Pieris brassicae* Boisduval and initial culture of parasitoids was started. Newly emerged males and females (Fig. 68) of *A. glomeratus* were caged in pairs in test tubes for mating. About 10-15 mated females were released into glass cage (type I Fig. 19a). Later 4th instar (Fig. 69) 20 larvae of *B. mori* (Pm × NB_4D_2) were introduced through the sleeve in the cage for quick oviposition. After parasitization (Fig. 63), the host larvae were placed in wooden trays for usual rearing by feeding M-5 mulberry (Fig. 2) leaves. Within 13 days the parasitoid larvae were emerged by breaking the body wall of silkworm and spun the cocoons (Fig. 67). Later, the cocoons were collected in test tubes for adult emergence. The cocoons were separated and kept in either specimen tubes or in test tubes. The mated females further exposed to the silkworm larvae for obtaining sufficient number of parasitized larvae. During the process of rearing, the parasitoids were fed with 50 per cent honey solution and *B. mori* larvae with mulberry leaves of M-5 variety.

3. Rearing of *A. angaleti*

The initial culture of parasitoid was maintained by collecting cocoons of *A. angaleti* from the fields either on Ain, *T. tomentosa* or Ber, *Zizyphus jujuba*. The field collected cocoons were reared separately either in individual specimen tubes (size 10 × 5 × 2) or in test tubes for adult emergence. Thus, newly emerged pair [1 male (Fig. 74) and 1 female (Fig. 73)] was caged in test tube (size 19 × 2.5 cm) for mating purpose. 10-15 mated females were exposed to 20 second instar *A. mylitta* larvae (Fig. 50, 79) in the insect cage type-II (Fig. 19b), for parasitization. After parasitization the parasitized larvae of *A. mylitta* were reared as described previously. Within 13.6 days the parasitoid larvae were found emerged from the parasitized tasar silkworms for spinning cocoons. These cocoons (Fig. 78) were later collected for adult emergence. During the rearing period the adult parasitoids were fed with 50 per cent honey solution and *A. mylitta* larvae with tender leaves of Ain trees (*T. tomentosa*).

III. HOST AGE SELECTION

Introduction

The silkworms *Bombyx mori* Linn. and *Antheraea mylitta* Drury are parasitized by several Hymernopterous and Dipterous parasitoids (Sathe, 1998). They attack different instars of the worms and cause threat to sericulture industry. Hence, the present work was carried out with three important parasitoid species namely *Exorista bombycis* Louis., *Apanteles glomeratus* Linn. and *Apanteles angaleti* Muesebeck.

There are more than 1/4 million species of parasitoids in the world, developing as larvae, feeding on the host insects, but differing enormously in behaviour and ecology.

Host age selection aspects are extremely important on the view point of the natural enemy management of the silkworms. Parsitoid development rate, adult size and adult longevity are related to the host stage attacked (Sathe and Ingawale, 1993a). Hence, parasitoids are classified into four categories by taking into account of the stage which they attack, i.e., egg, larval, pupal and adult parasitoids (Doutt, 1959). All categories of parasitoids show their preference for either younger or older developmental stage of the host (Waage and Greathead, 1986). The preferences of the parasitoids to the particular age of their host are related to hormones or alteration in the factors or alternations necessary for their acceptance (Smilowitz, 1974).

Jowyk and Smilowitz (1978) says that the host instar preferences of parasitoids may strongly influence the interaction between host and parasitoids. Such interactions have direct impact on introduction, colonization and mass production of parasitoids and are important aspects for planning biological control strategies. Host mortality rate also often depends on the instar (Hogg and Nordheim, 1983). Further more, host finding rate and thus potentially the amount of damage done to host plants may be affected by the instar parasitized. However, the host instar and species preferences of parasitoids have rarely been assessed with all instars and species of hosts with respect to host plants (Beckage and Riddiford, 1978). Force (1975) says that by attacking different stages, parasitoid partition the resource and that interspecific competition in one of the selective force increase parasitoid complexes. In many parasitoids, the function of preference is less clear or not known (van Alphen, 1980). The mature stages of hosts may be invulnerable to parasitoid attack, since they have more developed defence systems (Luck *et al.*, 1982).

In the present topic *E. bombycis* and *A. glomeratus* the parasitoids of *B. mori* and *A. angaleti*, a parasitoid of

PLATE VIII

Fig. 55. *E. bombycis* ovipositing on *B. mori* larva; **Fig. 56.** Eggs of *E. bombycis* on *B. mori* larva; **Fig. 57.** Scar by *E. bombycis* on *B. mori* larva; **Fig. 58.** First instar maggot of *E. Bombycis*; **Fig. 59.** Second instar maggot of *E. bombycis*; **Fig. 60.** Third instar maggot of *E. bombycis*; **Fig. 61.** Pupa of *E. bombycis* and **Fig. 62.** *E. bombycis* examining *B. mori* larva.

PLATE - VIII

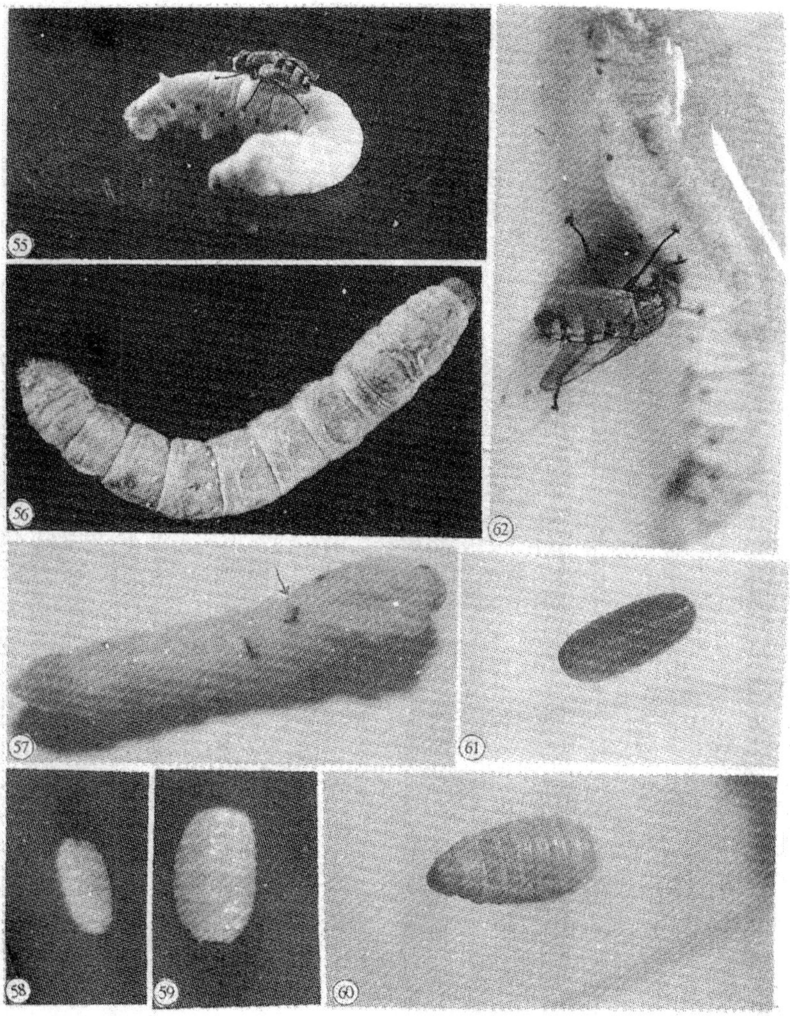

PLATE IX

Fig. 63. *A. glomeratus* ovipositing on *B. mori* larva; **Fig. 64.** Eggs of *A. glomeratus*; **Fig. 65.** First instar larva of *A. glomeratus*; **Fig. 66.** Third instar larva of *A. glomeratus*; **Fig. 67.** Cocoon colony of *A. glomeratus*; **Fig. 68.** Adult of *A. glomeratus*; **Fig. 69.** Rejection of host by *A. glomeratus*; **Fig. 70.** *A. glomeratus* watching for approach towards host. **Fig. 71.** *A. glomeratus* examining host.

PLATE - IX

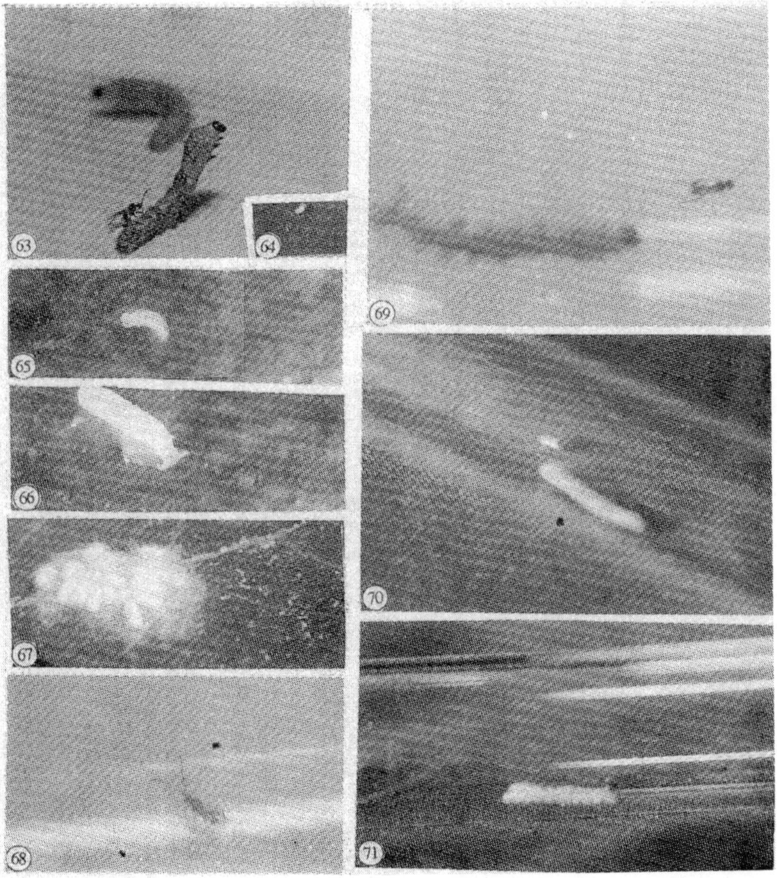

PLATE X

Fig. 72. *A. angaleti* female; **Fig. 73.** Egg of *A. angaleti*; **Fig. 74.** *A. angaleti* male; **Fig. 75.** First instar larva of *A. angaleti*; **Fig. 76.** Second instar larva of *A. angaleti*; **Fig. 77.** Third instar larva of *A. angaleti*; **Fig. 78.** Cocoon colony and adults of *A. angaleti*; **Fig. 79.** *A. angaleti* watching for approach towards host.

PLATE - X

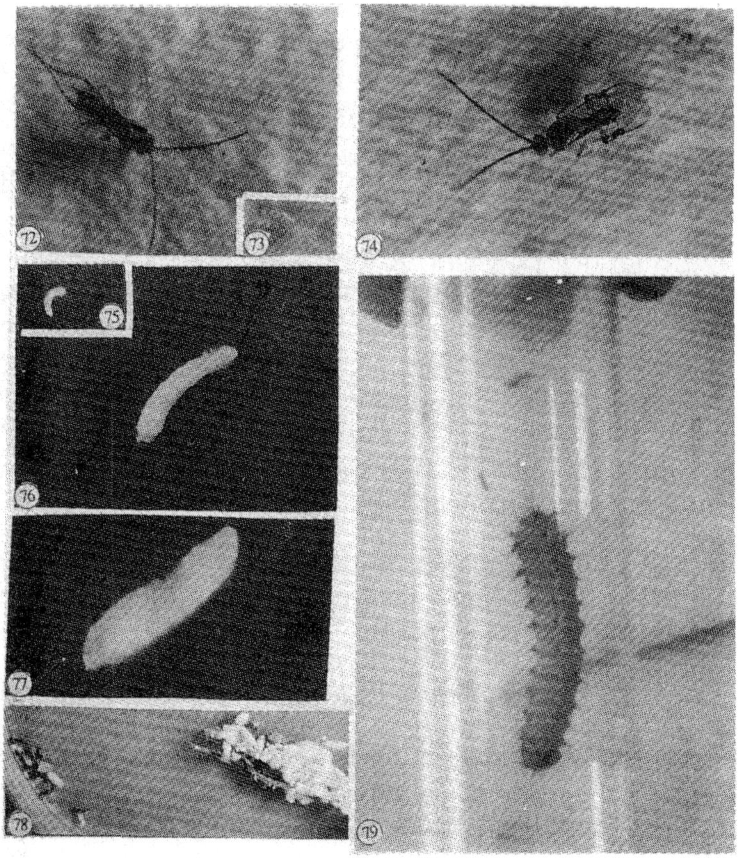

A. mylitta were selected for the studies. The uzi fly, *E. bombycis* is studied by several workers with respect to host age selection (Thompson, 1944; Isarangkul *et al.*, 1971; Ayuthaya, 1972; Datta and Mukherjee, 1978; Kumar *et al.*, 1990a; Kumar and Jolly, 1986a, b; Sood *et al.*, 1995; Manjunath and Puttaraju, 1996; etc.) However, on *A. glomeratus,* a gregarious parasitoid of *B. mori* there is a scanty literature and a very little attention is paid on this parasitoid. Chatterjee and Swarup, (1961) and Sood *et al.* (1995) reported this parasitoid from India on *Pieris brassicae* Linn. At present, *A. glomeratus* is not troublesome to sericulture industry, but it may pose threat to rearing *B. mori* worms as it is small sized and has some alternative hosts (Sood *et al.*, 1995). Knowledge on the aspects of host age selection by the parasitoid will add relevance for their management. Sood *et al.*, (1995) have studied this parasitoid with respect to age of *P. brassicae.* Its seasonal activity on *P. brassicae* have also been studied (Sood *et al.*, 1995).

A. angaleti is very important natural enemy of *A. mylitta* which cause 40 per cent mortalities on tasar larvae (Kole and Chatterjee, 1995).

In past, host age selection in parasitoids have been studied by Sathe (1984-1985), Chatterjee and Swarup (1961), Kumar *et al.* (1990b), Mahadevappa (1992), Sinchaisir *et al.* (1972), Sathe and Ingawale (1993a, b), Narsimha *et al.* (1993), Ram Kishore *et al.* (1993), Jyothi and Veeranna (1993), Savanurmath and Patil (1993), Thompson (1954), Prasad *et al.* (1985) etc. Hence, the objective of the present study is to find out the optimum age of silkworms for maximum parasitization.

MATERIAL AND METHODS

The silkworms were reared on the mulberry variety M-5 at laboratory conditions. (25±1 °C, 75 ± 3% R.H. and 16:8 hr L:D).

The culture of silkworms and prasitoids were maintained in laboratory. The silkworms of one day old to 26 day old were exposed to *A. glomeratus/E. bombycis* species for their parasitization in rearing cages (I and II Figs. 19a, 19b) with a constant host density 50 for *E. bombycis* and 20 for *A. glomeratus* for 24 hr. The parasitised larvae were reared in the laboratory for further observations of parasitoid emergence. Later, per cent parasitism was calculated. The experiment was replicated for 5 times. In case of *A. angaleti,* in door and out door experiments were conducted as described previously.

Results

1. Host age selection by *E. bombycis*

The results recorded in Table 16 and Fig. 80 showed that 100 per cent parasitism was seen on 17 to 26 day old host larvae. Minimum parasitism was noted on 4 day old larvae and it was found increasing up to 17 day old but, in the range of 17 to 26 day old larvae, there was not any significant difference in per cent parasitism. It was also noted that 6-day old larvae were not ideal for parasitism by uzi-fly wherein percentage parasitism ranged from 5.2 per cent to 16.8 per cent. Emergence of parasitoid was not seen in the larvae ranging from 1 to 3 day old.

2. Host age selection by *A. glomeratus*

Adults of *A. glomeratus* were emerged maximum (6%), from the host larvae which were 12-day old. The larvae ranging from 1-3-day old remain unparasitized. The results are tabulated in Table 17 and Fig. 81. The larvae of *B. mori* ranging from 4 to 21-day old were susceptible for pest attack. The sex ratio was in favour of females (1:1.01).

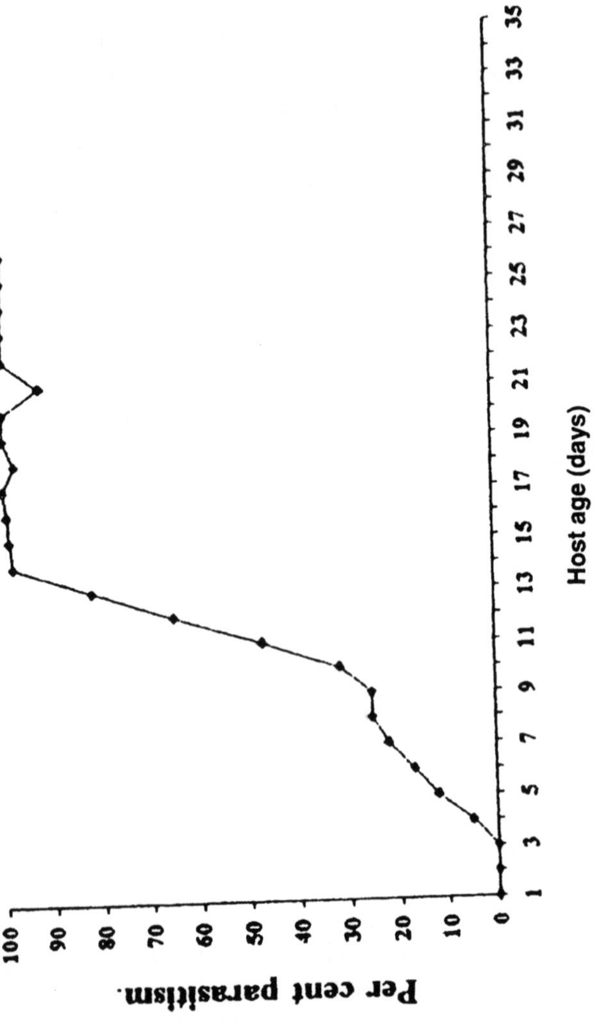

Fig. 80 : Relationship between Host age and per cent Parasitism by *E. bombycis*

Table 16 : Maximum Effective Age of *B. mori* Larvae for Parasitism by *E. bombycis*

Host age days	Total No. of hosts tried	Total No. moths emerged	No. of larvae parasitized	Total No. of parasitoids emerged			Sex Ratio			Per cent parasitism	Other mortality
				Male	Female	Total	Male	:	Female		
1	250	247	0	0	0	0	0	:	0	0	3
2	250	248	0	0	0	0	0	:	0	0	2
3	250	249	0	0	0	0	0	:	0	0	1
4	250	233	13	6	7	13	1	:	1.16	5.2	4
5	250	218	30	16	14	30	1	:	0.87	12.0	2
6	250	204	42	20	22	42	1	:	1.10	16.8	4
7	250	194	56	30	25	55	1	:	0.83	22.0	0
8	250	185	63	28	35	63	1	:	1.25	25.2	0
9	250	185	63	33	30	63	1	:	0.90	25.2	0
10	250	171	79	40	38	78	1	:	0.95	31.6	0
11	250	132	118	61	57	118	1	:	0.93	47.2	0
12	250	86	164	83	80	163	1	:	0.96	65.2	0
13	250	40	207	104	101	205	1	:	0.97	82.0	3
14	250	0	250	126	124	250	1	:	0.98	98.0	0
15	250	0	245	129	116	245	1	:	0.89	98.8	5
16	250	0	247	119	128	247	1	:	1.07	99.2	3
17	250	0	250	127	123	250	1	:	0.96	100.0	0
18	250	0	244	123	120	243	1	:	0.97	97.6	6
19	250	0	250	128	122	250	1	:	0.95	100.0	0

Contd.

Host age days	Total No. of hosts tried	Total No. moths emerged	No. of larvae parasitized	Total No. of Parasitoids emerged			Sex Ratio			Per cent parasitism	Other mortality
				Male	Female	Total	Male	:	Female		
20	250	0	250	124	126	250	1	:	1.01	100.0	0
21	250	0	243	122	121	243	1	:	0.99	97.2	7
22	250	0	250	126	124	250	1	:	0.98	100.0	0
23	250	0	250	124	126	250	1	:	1.01	100.0	0
24	250	0	250	127	123	250	1	:	0.96	100.0	0
25	250	0	250	124	126	247	1	:	1.01	100.0	0
26	250	0	250	127	123	250	1	:	0.96	100.0	0

Each host age group was replicated five times, each replicate consisted 50 larvae *of B. mori* and exposed for a period of 24 hr. to a mated female parasitoid.

Table 17: Maximum Effective Age of *B. mori* Larva for Parasitism by *A. glomeratus*

Host age days	Total No. of hosts tried	Total No. moths emerged	No. of larvae parasitized	Total No. of Parasitoids emerged			Sex Ratio	Per cent parasitism	Other mortality
				Male	Female	Total	Male : Female		
1	100	92	0	0	0	0	0 : 0	0	8
2	100	90	0	0	0	0	0 : 0	0	10
3	100	99	0	0	0	0	0 : 0	0	1
4	100	96	1	21	14	35	1 : 0.66	1	3
5	100	90	4	105	107	212	1 : 1.01	4	6
6	100	89	6	114	120	234	1 : 1.05	6	5
7	100	90	6	150	151	301	1 : 1.006	6	4
8	100	86	6	125	135	260	1 : 1.08	6	8
9	100	91	7	142	143	285	1 : 1.007	7	2
10	100	81	9	187	191	378	1 : 1.02	9	10
11	100	84	7	184	185	369	1 : 1.005	7	9
12	100	84	10	225	227	452	1 : 1.008	10	6
13	100	86	6	144	145	289	1 : 1.006	6	8
14	100	91	9	199	187	377	1 : 0.93	9	0
15	100	89	8	159	161	320	1 : 1.01	8	3
16	100	91	7	160	125	285	1 : 0.78	7	2
17	100	89	5	109	109	218	1 : 1.00	5	6
18	100	88	4	53	51	104	1 : 0.96	4	8
19	100	87	3	72	78	150	1 : 1.08	3	10

Contd.

Host age days	Total No. of hosts tried	Total No. moths emerged	No. of larvae parasitized	Total No. of Parasitoids emerged			Sex Ratio		Per cent parasitism	Other mortality
				Male	Female	Total	Male	: Female		
20	100	94	2	42	44	86	1	: 1.04	2	4
21	100	92	2	38	38	76	1	: 1.00	2	6
22	100	92	0	0	0	0	0	: 0	0	0
23	100	91	0	0	0	0	0	: 0	0	0
24	100	94	0	0	0	0	0	: 0	0	0
25	100	90	0	0	0	0	0	: 0	0	0
26	100	91	0	0	0	0	0	: 0	0	0

Each host age group were replicated five times, each replicate consisted 20 larvae of *B. mori* and exposed for a period of 24 hr to a mated female parasitoid.

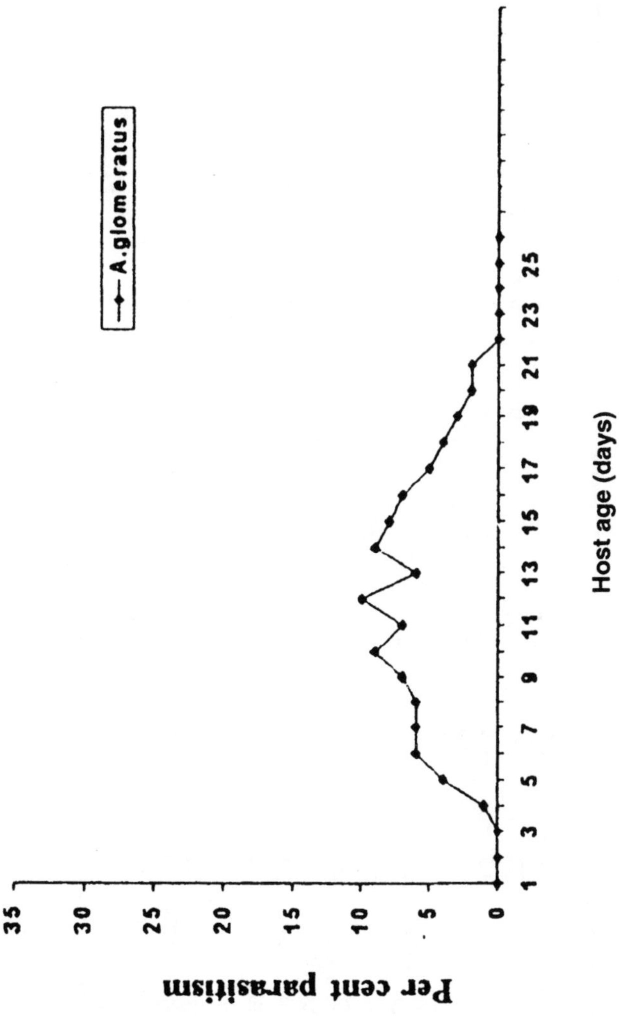

Fig. 81 : Relationship between host age and per cent parasitism by *A. glomeratus*

Table 18 : Maximum Effective Age of *A. mylitta* Larva for Parasitism by *A. angaleti*

Host age days	Total No. of hosts tried	Total No. moths emerged	No. of larvae parasitized	Total No. of Parasitoids emerged			Sex Ratio			Per cent parasitism	Other mortality
				Male	Female	Total	Male	:	Female		
1	50	20	0	0	0	0	0	:	0	0	30
2	50	21	0	0	0	0	0	:	0	0	29
3	50	23	0	0	0	0	0	:	0	0	27
4	50	19	1	27	29	57	1	:	1.07	2	30
5	50	22	2	58	56	114	1	:	0.96	4	26
6	50	18	3	64	67	131	1	:	1.04	6	29
7	50	17	6	73	75	148	1	:	1.02	12	27
8	50	19	8	68	64	132	1	:	0.94	16	23
9	50	20	12	35	33	68	1	:	0.94	24	18
10	50	15	15	267	271	538	1	:	1.01	30	20
11	50	21	9	84	85	169	1	:	1.01	18	20
12	50	21	8	79	72	142	1	:	1.02	16	21
13	50	18	6	19	20	39	1	:	1.05	12	26
14	50	22	8	57	57	114	1	:	1.00	16	20
15	50	23	2	56	56	112	1	:	1.00	4	25
16	50	12	4	103	107	210	1	:	1.03	8	34
17	50	02	4	97	95	192	1	:	0.97	8	44
18	50	12	2	43	45	98	1	:	1.04	4	36
19	50	8	3	44	46	90	1	:	1.04	6	39

20	50	14	2	148	151	299	1 : 1.02	4	34
21	50	09	2	53	49	0.2	1 : 0.92	4	39
22	50	12	1	21	30	51	1 : 1.42	2	37
23	50	8	9	0	0	0	0 : 0	0	42
24	50	9	0	0	0	0	0 : 0	0	4
25	50	11	2	0	0	0	0 : 0	0	39
26	50	14	0	0	0	0	0 : 0	0	36
27	50	2	0	0	0	0	0 : 0	0	48
28	50	11	1	0	0	0	0 : 0	0	39
29	50	19	2	0	0	0	0 : 0	0	31
30	50	7	3	0	0	0	0 : 0	0	43
31	50	14	6	0	0	0	0 : 0	0	36
32	50	13	8	0	0	0	0 : 0	0	37
33	50	19	12	0	0	0	0 : 0	0	31
34	50	4	10	0	0	0	0 : 0	0	46
35	50	16	9	0	0	0	0 : 0	0	34

Each host age group was replicated five times, each replicate consisted 10 larvare of *A. mylitta* and exposed for a period of 24 hr to a mated female parasitoid.

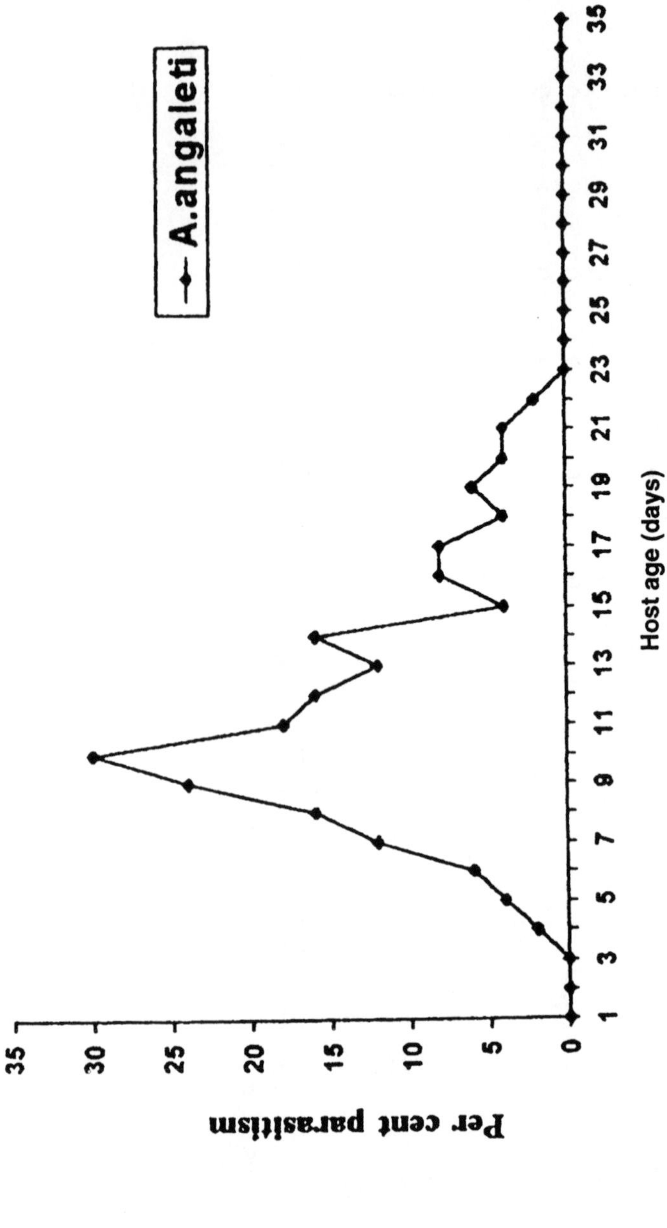

Fig. 82 : Relationship between host age and per cent parasitism by *A. angaleti*

3. Host age selection by *A. angaleti*

The results are recorded in Table 18 and Fig. 82. The maximum parasitism (30%) was noticed on 10 day old hosts, 1-3 day old larvae were not accepted for parasitism by *A. angaleti*. Higher number of parasitoids were emerged from the larvae which were 9-17 day old. The host larvae, ranging from 5 and 17 day old were readily accepted by the parasitoids. The sex ratio was in favour of females. The larvae ranging from 4 to 22 day old were susceptible for pest attack.

Discussion

The size and age are usually related to the sex ratio (Vinson, 1976). The host size was found to influence host choice (Recherson and Deloach, 1972). The host size preferences for certain immature hosts have been reported by Miller (1970). Age dependent size was also responsible to influence the host acceptance (Gutierrex, 1970). The relationship between host age and rate of parasitization was investigated by Leong and Oatman (1968).

The uzi flies of silkworms have been attempted by several workers (Sinchaisir *et al.*, 1972; Prasad *et al.*, 1985; Kumar *et al*, 1990; Narsimha *et al.*, 1993; Ramkishore *et al.*, 1993; Savanurmath and Patil, 1993 etc.) with respect to host age selection.

Sinchaisir *et al.* (1972) have studied the host preference of the tachinid fly to different stages and varieties of silkworms and they concluded that *Tachina sorbillans* preferes to lay their eggs on the later stages and Japanese varieties to younger stages of silkworm, probably due to larger body size.

Prasad *et al.* (1985) studied the physiological interactions between *B. mori* and its endoparasitoid *E. sorbillans,* they reported that the mortality was high in single and multiple uzi infections when the silkworm of Hosa Mysore (HM) and Kalimponga-A (KA) races were infected early in their larval life viz., 1st and 2nd instars. When older worms were infected, the mortality in larval stages reduced; the mortality in such cases were dependent on the acuity of infection. A proportional increase in mortality commensurate with the number of live maggots in the body of silkworm, was not observed since the maggots in multiple infection had stunted growth rates. Mortality in young larva of KA-U (KA infected with uzi maggots) was higher than that of HM-U (HM infected uzi maggots) whereas, older larvae of KA-U showed lower mortality than those of HM-U. The difference in the larval weight in the 4 grades of infection was very large in young larvae and less in older larvae. Infected younger larvae of HM survived for longer duration than KA larvae of the same age group whereas, one day worms stabilized to uzi infection in KA, it could not be observed upto 17 days of larval infection in the case of HM.

The observations made by Narasimha Rao *et al.* (1993) on *E. sorbillans,* a parasitoid of *B. mori* reveal that uzi-fly parasitized the developmental stages of silkworm including chawki worms, except 1st instar. The infestation of the fly was severe with the advancement of development (Chawki to 5th instar) confirming to view of Siddappaji and Basavanna (1990) that ovipositing flies preferred older larvae. These observations also lead a strong support to the suggestions of Kasturi bai *et al.* (1986) that 4th and 5th instar silkworms produce a Kairomone which is strongly attractive to the uzi fly. The instar wise per cent infestation was 6.59, 6.16, 9.35 and 16.03 per cent for 2nd, 3rd, 4th and 5th instars respectively. Jyothi and Veeranna (1993) reported that in *E. sorbillans,* the host pupae exposed for parasitization at ages ranging from 4-5 day old were suitable for infestation.

Savanurmath and Patil (1993) studied the host stage preference in tasar uzi fly, *Blepharepa zebina* Walker and reported that the tasar uzi fly *B. zebina* laid 10 times more eggs on the host of 4th and 5th instar as compared with those laid on hosts of other instars. Accordingly, 1st and 2nd instar worms were among the least preferred. Quite interestingly, no eggs were laid on worms which were either under moult or ready to spin (Savanurmath and Patil, 1993). Like wise, least (on an average 1.67 per larva) number of eggs were laid on pebrinized hosts. Moreover, it was noticed that even among the worms of preferred instars, the fly selected the well fed worms as against the poorly developed ones. In present study, observations were made on the host age selection by *E. bomcycis* and absolute parasitization was occurred on mulberry silkworms ranging from 17 to 26 days old and 1 to 3 day old larvae remain unparasitized by *E. bombycis*. However, a larvae ranging from 4 day old to 26 day old were susceptible for pest attack. The sex ratio was found favouring the males on the older instars than that of the younger instars, these results are not in agreement with the previous views of Charnov *et al* (1981). *E. bombycis* is thoroughly studied with respect to host age and parasitism previously hence, the present study was aimed to confirm the results.

Thangavelu and Singh, (1993) reported that the female parasitized the tasar silkworm larvae from the 3rd stage onward. The gravid female oviposited the silkworm and the maggot enters the body of the silkworm and feed inside the silkworm larvae. The infested silkworm larvae can be identified from a distance by the presence of black spot on the larval body due to the maggot entry. The infested larvae form flimsy cocoons and ultimately the tasar pupa dies.

The tachinid or uzi flies belongs to order Diptera, superficially resemble with house flies and flesh flies which cause the damage to their hosts by parasitizing them. In general the tachinid flies are grouped into two, i.e. macrotype and microtype. The individuals which have the macrotype of eggs, they lay their eggs on the host body, while the microtype forms deposit their eggs on the food material of the host. The Indian uzi, *E. bombycis* comes under macrotype egg form, while, the Japanese uzi, *Crossocosmia sericariae* Rondani comes under microtype egg form and deposit their eggs on the mulberry leaves along the leaf vein thus, through food, the worms get infected.

The host age preference by *Campoletis chlorideae* Uchida (Ichneumonidae) towards four lepidopterous host species viz. *Pseudoletia unipuncta* (Howorth), *Trichoplusia ni* (Hubner), *Prodenia eridinia* (Cramer) and *Pseudoletia praefica* Grote was investigated by Lingren *et al.* (1970). The larvae of 1 day old of all the hosts were susceptible for parasitism, 2 to 6 day old were readily parasitized and 2 to 4 day old being most acceptable. Many workers (Hays and Vinson, 1971; Lewis, 1971, Smilowitz, 1974) reported that as a larval host reaches the pharate pupal state, it become unacceptable for parasitism. Changes in the host acceptance have been related to hormones (Smilowitz, 1974) or to alteration in the factors necessary for acceptance. Thus, in most studies the importance of size and age remained in question, until the chemicals involved.

In *Apanteles (=Pseudapanteles) dignus* (Muesebeck) (Braconidae), a parasitoid of *Keiferia lycopersicella* (Walsigham) maximum parasitism was recorded on 2 or 3 day old host larvae and the highest number of parasitoids per replicate (12) was obtained. Decrease in the progeny production was observed gradually with an increase in the age of the host larvae beyond 2-3 day old. No parasitoids were obtained when larvae of 9 day old were exposed (Cardona and Oatman, 1971).

In *Agathis gibbosa* (Say) (Braconidae), Odebiyi and Oatman (1972) reported that 2-3 day old larvae were more suitable for parasitization. The highest per cent parasitization, 52.95 and the number of parasitoids per replicate, 19.3 per cent were obtained with this host age group. They also observed that progeny production decreased with an increase in age of host larvae beyond 2-3 day old, specially in those older than 4-5 days. In the present study in *A. glomeratus* (Braconidae), it is observed that, maximum parasitization was noted on 12 day old *B. mori* larvae on which maximum number of parasitoids have been obtained, 1-3 day old larvae were rejected by the parasitoid for parasitism. The larvae ranging from 5 to 11 day old were readily accepted for parasitization. However, beyond the host age 12 days, the per cent parasitism and progeny production decreased (upto 21 days), while, 22-26 day old larvae were rejected for parasitization by the parasitoid.

Oatman and Platner, (1974) showed that 3-4 day age of the caterpillars of *Phthorimaea opercellella* (Zeller) was most suitable for maximum parasitization by *Temelucha* sp. of *platensis* group. They also noted that *Hyposoter exoguae* (Viereck), a solitary endoparasitoid of the cabbage looper, *T. ni* preferred to oviposit in young host larvae and showed greatest preference for 1st instars (Smilowitz and Iwanstsch, 1975). A maximum of 31 per cent parasitism was noted with 1st instars and decline in per cent parasitism was noted in further instars. While, in the present gregarious braconoid form (*A. angaleti*) maximum parasitization was occurred on 10 day old *A. mylitta* larvae.

In some parasitoids it is seen that parasitoids kill their hosts by feeding on them and oviposit in others. The apheinid, *Encarsia formosa* oviposits in white fly larvae of the 3rd and 4th instar and prepupae, but feeds mainly on

2nd instars and pupae (van Alphen *et al.*, 1976), van Alphen (1980) found that the eulophid *Tetrastichus asparagi* Crawford used young eggs of host for feeding and oviposited in eggs with a developed embryo. The phenomena of feeding on host eggs by female parasitoids was not observed with the present forms viz. *E. bombycis, A. glomeratus* and *A. angaleti*. However, some Icheumonid parasitoids like *C. chlorideae,* a parasitoid of Gram pod borer, *Helicoverpa armigera* (Hubn.) and *Eriborus argenteopilosus* (Cameron) feed on host haemolymph ozed from the host at the time of oviposition (Sathe and Santhakumar, 1988a; Sathe, 1990b). In the present study such type of feeding behaviour was not observed in the parasitoids.

It was observed that in some cases the function of preference is less clear (van Alphen, 1980). In *Tetrastichus asparagi* Crowford handling time of older hosts was much longer because of defence by the host, female oviposited normally in eggs but, attacked larvae when eggs were scarce (van Alphen, 1980). The probability of encounter with older and therefore, larger hosts may be higher than with younger hosts (van Alphen and Drijwar, 1982) and may counteract the prolonged handling time.

The development rate of a parasitoid is dependent on the host instar parasitized (Thurston and Postley, 1978). In *Apanteles (=Cotesia) congregatus* (Say), a parasitoid of *Manduca sexta* (L.) observed that the development of the parasitoid was slower when oviposition occurred in 1st or 2nd instars than occurred in 3rd or 4th. In present form particularly in *A. glomeratus* and *A. angaleti* which are gregarious, maximum parasitization was seen on 12 and 10 day old host larvae respectively. Leibee *et al.* (1979) reported optimum parasitization by *Potsson lameerei* Debauche (Hymenoptera : Mymaridae) on 2 day old *Sitona hispidulus* (Fabricius) larvae. They further reported that females have difficulty in inserting their ovipositor into older eggs. Therefore, it was suggested that the reduced parasitism after the optimum age was due to the harder chorion of the older eggs. The present parasitoid forms are larval parasitoids hence, the above phenomenon was not observed.

The reproductive potential of *Apanteles creatinoti* Viereck was studied in relation to age of caterpillar of *Thiocidas postica* Wlk. by Sathe *et al.* (1987a) and found maximum parasitism, 41.3% with 4-5 day old hosts. Recently, Sathe and Ingawale (1993a) studied the host age selection in *Apanteles blateatae* Lal, a parasitoid of *Sylepta derogata* (Fab.) and found that maximum adult emergence, 48 per cent, was on 4 day old hosts. Host larvae older than 6 days have been progressively less suitable. The caterpillars of 3 to 6 day old were readily accepted by parasitoid and yielded higher number of parasitoids. Similarly, Sathe (1993) studied host age selection by *Apanteles jayangarenis* Bhatnagar (Hymenoptera, Braconidae), a larval parasitoid of *Spilosoma obliqua* (Wlk.). He also reported maximum parasitism (43%) on 4 day old hosts. Host larvae older than 4 days have been progressively less suitable. The caterpillars of 3 to 5 day old were readily accepted by parasitoid and yielded higher number of parasitoids.

In *Apanteles (Cotesia) orientalis* C. and N., a solitary parasitoid of *Exelastis atomosa* (Walsingham) parasitization was not seen in 1 and 8 to 10 day old host larvae. Caterpillars of 2 to 8 days age were susceptible for parasitization. Maximum (31.5%) parasitization was recorded on 3-4 day old larvae. There was significant ($P<0.05$) correlationship between age of host and per cent parasitism (Sathe, 1988b), while, the gregarious wasp *Apanteles (Cotesia) flavipes* (Cameron), a parasitoid of Jowar stem borer, *Chilo partellus* (Swinhoe) preferred older (7-8 day old) host larvae for parasitization, on which yielded maximum, 42 per cent parasitism (Nikam and Sathe, 1983b). Nair (1988) showed that the same wasp successfully parasitized the diapausing larvae of *C. partellus*. In present forms, perticularly in *A. glomeratus* and *A. angaleti* which are gregarious, maximum parasitization was seen on 12 and 10 day old host larvae respectively. The above findings showed that the solitary parasitoid prefers younger hosts whereas, the gregarious older. The reasons in solitary parasitoids in doing so would be (i) generally less/one, eggs/egg, are/is laid in a single host and only one parasitoid develops among them, (ii) the nutrient requirement for the development of single parasitoid is less, iii) while depositing their eggs, the females have to face the resistance exhibited by older hosts, as the hosts become more defendant. While, in gregarious forms ; (i) number of parasitoids have to develop from a single host (ii) the parasitoids require relatively large quantity of food (iii) probably the hosts might be less defendant.

Several species of parasitic wasps discriminate hosts of different size (Kishi, 1970; Sandlan, 1979; Jones *et al.*, 1982; van den Assem *et al.*, 1984). The small hosts were preferably used for male offspring and large hosts for female (Charnov *et al.*, 1981) but, in *Lariophagus distinguendus*, a solitary, larval parasitoid of *Sitophilus grannarius* very few females emerged from larger hosts and hardly any females from larger and intermediate size host, indicating that the host size was an important factor in the chain of process which lead to sex determination (Simboloti *et al.*, 1987). In the present study *A. angaleti* showed the sex ratio favouring the females in older instars. The findings are in agreement with the opinion of Charnov *et al.* (1981). However, the results obtained with *A. glomeratus* are not in agreement with the opinion of Charnov *et al.* (1981) and in *E. bombycis* there is no correlation with host age, size and sex ratio. In general present findings on host age selection, in these three parasitoids, may be helpful for designing appropriate control measures of the flies.

IV. HOST SPECIFICITY

Introduction

In the entomophagous insects an analysis of host specificity discloses many factors that determine whether or not any two given species are to be associated as host and parasitoids. There are some species, which limit their attack to a single host while others found in nature to attack several different hosts. None of the parasitoids appears to be completely indiscriminable. Under natural conditions they will attack only a fraction of species on which development is possible. Under laboratory conditions, one can remove the barriers of time and space which separate host and parasitoids in nature. The breeding of parasitoids in the laboratory on unnatural hosts has a great importance in large scale rearing of parasitoids for field colonization. Certain species of parasitoids are more amenable to laboratory production than others and their mass culture is simplified on factitious hosts and can often be put on commercial basis. Salt (1937) says that "an instance in which the parasitoid actually chooses its own species in preference to the ancestral host, has yet to be described among entomophagous forms." Obviously, a parasitoid limit its attacks to a fraction of suitable host species that occur in nature. This stimulates the investigator to study the behaviour of the adult parasitoids and the criteria used by the females for selecting their host species.

Salt (1935) reported four distinct and consecutive steps of host selection by parasitoids. The four steps are habitat selection, host location, host acceptance and host suitability.

Habitat Selection

An entomophagous insect that emerges in a habitat not populated by its host have the capacity to locate the habitat in which its host occurs. Thus, habitat selection is a very important 1st step in the host selection sequence of events. The movement of parasitoids i.e. outside the habitat of its host must be either random or be influenced by some environmental characters or characters typical for that habitat. A parasitoid initially and fundamentally seeks a certain environment and it does this irrespective of the presence or absence of host.

Host Finding

Once a parasitoid has located an appropriate habitat, it must be able to find a suitable host. In this second stage of host selection, chemicals released from the host play an important role (Vinson, 1980). Much more is known about this process than about the habitat selection (Persoons *et al.*, 1993).

Host Acceptance

Once the parasitoid has located its host by means of airborne volatiles acting over relatively long distances, a new group of semiochemicals start playing a role which induce oviposition or larvae position, some times the same compounds that attract the parasitiod from a distance, but in other cases different compounds are responsible. When a parasitoid finds or contacts suitable host, it still may not attack the host if the proper stimuli are lacking. This is the third step of host selection pertaining to host specificity of the parasitiod.

Host Suitability

Though the parasitoid has found the host in its habitat and selected it for attack, it may still not succeed if the potential host individual is immune or otherwise unsuitable. This is the fourth and final step in the process by which the host of a parasitoid species becomes restricted.

The host preferences by parasitoid(s) may vary depending upon host species and/or host plant(s) (Weseloh, 1981). The contamination of the odour of a preferred host may result in the unacceptability of the host (Thrope and Jones, 1937). Vinson (1975) says that some non host species are readily attacked when encountered. Several workers (Gutierrez, 1970; Lingren *et al.*, 1970; Lingren and Noble, 1972; Lewis and Vinson, 1971; Drooz and Fedde, 1972; Calvert 1973; Jackson *et al.*, 1979 and Hopper and King, 1984; etc.) studied the host specificity and host parasitoid interactions in certain models of insects.

In the present study following species namely, *Bombyx mori* Linn., *Artheraea mylitta* Drury., *Spodoptera litura* Fab., *Agrotis ypsilon* (Rott.) and *Pieris brassicae* Linn. were considered as host for the parasitoids, *Apaneteles glomeratus* Linn., *Apanteles angaleti* Muesebeck, and *Exorista bombycis* Louis. Host specificity is studied in *E. bombycis* in India by several workers (Narayanswamy *et al.*, 1993; Beeson and Chatterjee, 1935; Siddappaji and Jamil, 1992; Kumar *et al.*, 1990b), but it is not studied in *A. glomeratus* and *A. angaleti*. Keeping in view the importance of host specificity in pest management, the present topic was selected.

Material and Methods

Laboratory cultures of silkworm and their parasitoids were used for the present study. To find out the host preference given by the parasitoid species, the hosts *B. mori, A. mylitta, A. ypsilon, S. litura* and *P. brassicae* were exposed in glass cage (type I, Fig. 19a) with mixed host density 100 (20 for each species) to a single mated female of *E. bombycis, A. glomeratus* and *A. angaleti* individually. After 24 hr exposure the larvae were separated in trays/throughs and reared for parasitoid/moth emergence. The parasitoids and hosts were fed with 50 per cent honey solution and host food plant leaves respectively. During the course of studies the mortality in silkworm species was less than 10 per cent. The experiments were conducted at laboratory (25±1 °C, 75±3, R.H., Photoperiod 18:6, hr L:D) and replicated for 5 times.

Results

Host Preference by *E. bombycis*

Except *P. brassicae* all exposed host species were parasitized with higher per cent parasition by *E. bombycis*. Maximum, 98 per cent parasitism was seen on *B. mori* (Pm × NB_4 D_2) while, it was 67 per cent on *A. ypsilon* and 88 per cent on *A. mylitta*. The parasitism performance by *E. bombycis* on *B. mori, A. ypsilon, S. litura, P. brassicae* and *A. mylitta* is recorded in Table 19. The per cent of moth emergence was highest in *P. brassicae* but, the percentage of moth emergence (Fig. 42) was less in *B. mori* (Pm × NB_4 D_2). The results are shown in Table 19 and Fig. 83.

Host Preference by *A. glomeratus*

The parasitoids get attracted towards all the host species exposed, *B. mori* (Pm × NB_4D_2), *S. litura, A. mylitta, A. ypsilon* and *P. brassicae*. *B. mori* (Pm × NB_4D_2) showed 55.00% parasitism while, *P. brassicae* showed maximum 60 per cent parasitism by *A. glomeratus*. The other hosts offered to the parasitoid remain unparasitized. The results are shown in Table 19 and Fig. 83.

Host Preference by *A. angaleti*

A. angaleti preferred the host species *A. ypsilon* with 53 per cent parasitism and *A. mylitta* with 47 per cent

Table 19 : Host Preference by the Parasitoids of Silkworms

Parasitoid species	Host species	Host density (mixed)	Per cent parasitism	Per cent host mortality	Per cent moth emergence
E. bombycis	*B. mori* (Pm x NB_4D_2)	100	98	2	0
	A. mylitta	100	88	10	2
	S. litura	100	73	9	18
	A. ypsilon	100	67	7	26
	P. brassicae	100	0	5	95
A. glomeratus	*B. mori* (Pm x NB_4D_2)	100	55	8	37
	A. mylitta	100	0	66	34
	S. litura	100	0	4	96
	A. ypsilon	100	0	5	95
	P. brassicae	100	60	7	33
A. angaleti	*B. mori* (Pm x NB_4D_2)	100	0	7	93
	A. mylitta	100	47	36	17
	S. litura	100	0	8	92
	A. ypsilon	100	53	8	39
	P. brassicae	100	0	6	94

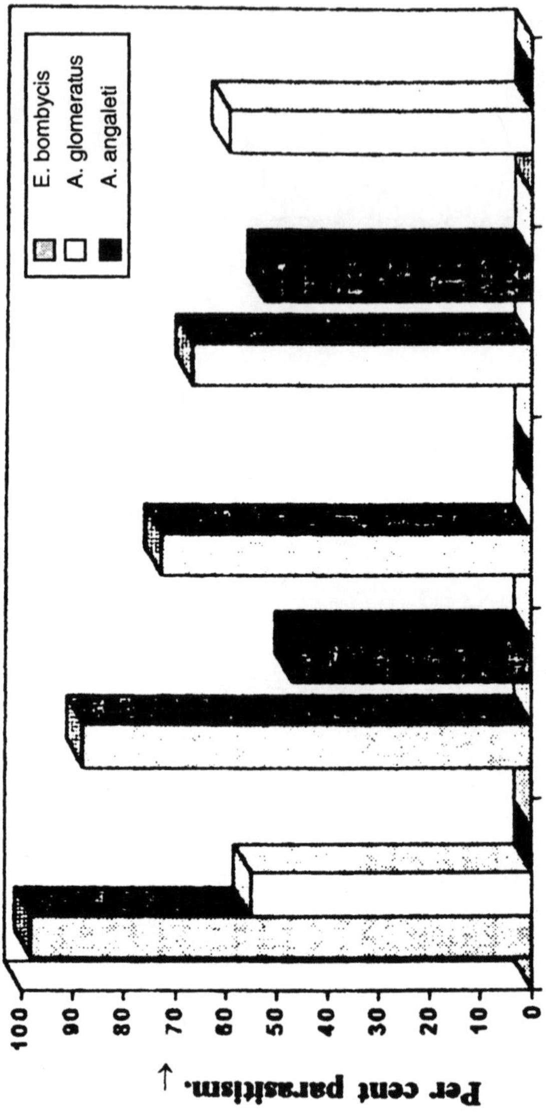

B.m = *B. mori*, A.M. = *A. mylitta*, S.l = *S. litura*, A.g = *A. ypsilon*, P.b = *P. brassicae*.

Fig. 83 : Host Preference by the Parasitoids

parasitism other host species, i.e. *B. mori* (Pm × NB_4D_2), *S. litura* and *P. brassicae* remain unparasitized. Per cent host mortalities and moth emergence in the exposed species are shown in Table 19 and Fig. 83.

DISCUSSION

Narayanswamy *et al.*, (1993) studied the host preference of *E. bombycis* by providing eleven species of Lepidopteran caterpillars. Their results showed that uzi fly preferred *B. mori* larvae (54.52%) for oviposition followed by *Samia cynthia ricini* Boisdual (13.22%), *A. mylitta* (10%), *S. litura* (8.9%), *Heliothis armigera* (Hubn.) (7.41%), *Achea janata* L. (5.16%) and *Adisura atkinsoni* Moore (0.86%). Further, they found that the fly did not lay eggs on caterpillars of *Spilosoma obliqua* (Walk.), *Euproctis fraterna* M. *Eupterote mollifera* W. and *Attera fabriciella*. These authors further reported that the incubation, maggot, prepupal, pupal and total developmental periods of uzi fly were considerably shorter besides increased watching, adult emergence and longevity of both male and female flies on. *B. mori* larvae followed by the *S. c. ricini, H. armigera, S. litura, A. mylitta* and *A. janata*. However, adult emergence was not seen from the puparia of *A. atkinsoni*. These authors also tried one more host, i.e. *Corcyra cephalonica* wherein egg hatching was successful but the larva died due to the death of *C. cephalonica*.

Narayanswamy *et al.* (1993) concluded that texture of the skin, physical obstructions and also body size of hosts seemed to have influenced the rate of oviposition of *E. bombycis* on the eleven species screened. They also concluded that the parasitoid perpetuates on secondary hosts in the absence of primary host, *B. mori*. In the present study 5 lepidopterous hosts were exposed viz. *B. mori, A. mylitta, S. litura, A. ypsilon* and *P. brassicae* and yielded

98 per cent, 88 per cent, 73 per cent, 67 per cent and 9 per cent adults of uzi fly respectively. The new host, *P. brassicae* remain unattacked by uzi fly. The probable reason may be the same as shown by Narayanswamy *et al.* (1993), i.e. host body skin structure, host size and other obstructions. However, in the present study, the per cent parasitism was much higher on *B. mori, A. mylitta* and *A. ypsilon* than reported earlier.

Beeson and Chattarjee (1935) reported that *Trycholyga bombycis* Beck. can attack several lepidopterous insects like *Acherontia lachesis* F. (Sphingidae), *Attacus ricini* Boisduval (Saturniidae), *B. mori* (Bombycidae), *Andraca bipumata,* Walk. (Bombycidae), *Dasychira mendosa* Hubner, *D. thwaitesti* Moore, *Thiocidas postica* Walk. (Lymentriidae) and *Metanastria hyrtaca* Cameron (Lasiocampidae). Crosskey (1977) published a catalogue of the Diptera of the oriental region of Tachinidae in which he maintioned the names of the tachinids known to attack lepidopterous hosts including those which produce silk.

Isarangkul *et al.* (1972) studied the host specificity in a tachinid fly, *Tachina sorbillans* from Bangladesh and reported that caster semi looper, *Achea janata* L. and cotton leaf worm, *Prodinia litura* Fab. were important hosts of *T. sorbillans.* They also noted that former was more favoured than the later. In the present study order of preference to host by *E. bombycis* was, *B. mori* > *A. mylitta*>*S. litura* > *A. ypsilon.* Sabrosky and Reardon (1976) reported a new host, Gypsymoth *Lymantria disper* L. for *T. sorbillans.* Similarly, Siddappaji and Jamil (1992) studied the host range of Indian uzi fly *E. bombycis* and reported 44 host species of different lepidopterans belonging to 36 genera including silkworms. Thompson (1951) recorded 44 species of Lepidopteran hosts to *T. (=E.) bombycis* belonging to 36 genera of families Sphingidae, Bombycidae, Amathusiidae, Arctiidae, Saturniidae, Lasiocampidae, Drepamidae,

Hesperiidae, Phalaenidae, Lymantridae, Psychidae, Nymphalidae, Lymacodidae and Thoumetapocidae. However, the recent record of hosts of *E. bombycis* is 81 hosts from Lepidoptera and one is from Hymenoptera (Anonymous, 1993).

Switching and optimal host selection are the two theoretical approaches which are applicable to host selection by parasitoids. The tendency of parasitoids to concentrate on the most abundant host species has been called switching (Murdoch, 1969). Switching depends on the preference of a parasitoid for one of the host species and preferences is usually measured as a deviation in the proportion of host attacked and the host available in the environment (Cornell, 1978). When a parasitoid has a choice between two host species it would accept the best one in terms of fitness, gain and that the second would only be accepted if it would increase number of offspring produced per time unit (Erulen, 1966; Charnov, 1976; Iwasa *et al.*, 1984). Thus, depending on encounter rate with the best host species, the less profitable species would either be accepted or rejected (Mc Arthur and Planka, 1966). In parasitic hymenoptera host specificity is studied by some workers (Tawfik, 1957; Jackson *et al.*, 1979; Sathe and Santhakumar, 1992a).

Lingren *et al.*, (1970) studied the host preference in an Ichneumonid, *Campoletis perdistinctus* Viereck (Hymenoptera) and reported host 27 host species of Lepidoptera. However, none of the hosts were well adopted to mass rearing as like greater wax moth and mediterranean flour moth. However, the alfalfa looper, boll worm, cabbage looper, fall armyworm, southern armyworm, variegated cutworm and tobacco bud worm were very susceptible for parasitism.

Later, in 1972, Lingren and Nobel reported host preference performed by *C. perdistinctus* towards noctuid larvae, cabbage looper, *Trichoplusia ni* (Hubner), beet armyworm, *Spodoptera exigua* (Hubner), boll worm, *Heliothis zea* (Boddle), fall armyworm, *Spodoptera frugiperda* (Smith), southern armyworm, *Prodenia eridania* (Cramer), tobacco budworm, *Heliothis virescens* (F.) and yellow stripped armyworm, *Prodenia ornithogalli* Guenee. Their results indicated that the bollworms, fall armyworms and tobacco budworms were mostly preferred in all laboratory tests, the cabbage looper was the least preferred and no cocoons were produced from the larvae of the beet armyworms. The results indicate that the bollworm and tobacco budworm were among the most preferred noctuid hosts in the laboratory as well as in the field tests.

Jackson *et al.* (1979) reported that the pink bollworm, *Pectinophora gossypiella* (Saunders), cotton bollworm, *H. zea,* tobacco budworm, *H. virescens,* cabbage looper. *T. ni* and beet armyworm, *S. exigua* were the most suitable hosts for *Chelonous (=Microchelonus) blackburnii* Cameron (Hymenoptera, braconidae) however, the saltmarch caterpillars, *Exstigmene acrea* (Drum.) exposed to above braconoid parasitoids were not suitable. Among exposed hosts, the order of preference was as follows: Pink bullworm > cabbage looper > cotton bollworm > tobacco budworm > beet armyworm. In the present study, for *A. glomeratus,* 5 hosts were exposed namely, *B. mori, A. mylitta, S. litura, A. ypsilon* and *P. brassicae.* Out of these, only two, *B. mori* and *P. brassicae* were parasitized with high per cent parasitism.

Recently, Sathe and Santhakumar (1992a) studied the host specificity in *Campoletis chlorideae* (Uchida), parasitoid of Gram pod borer *Helicoverpa armigera* (Hubner). Their results showed that the parasitoid preferred 3 lepidopterous species namely *H. armigera, S. litura* and *S. exigua* while, *Exelastis atomosa* Walsingham larvae were rejected.

Hopper and King (1984) showed the host preference of *Microplitis croceipes* (Cresson) towards *H. zea* and *H. virescens* larvae. The wasps preferred 3rd instars most and the preference was not dependent on total density or on the density of the most preferred instars. Further, the sex ratio of the wasps was not affected significantly by the instar parasitized and the development time in an instar was also universally correlated to the observed pattern of preferences. In the present study, larvae exposed to parasitoid were of 4 day old and the density of the hosts were constant, i.e. 100 (25 of each species) and the sex ratio was in favour of females. In the present form, *A. glomeratus* the host density was 100 of mixed culture of 5 hosts and exposed for 24 hr.

Azuma and Kitano (1971) studied the host selection test in *A. glomeratus.* In each of five cages (70 × 60 cm nylon net cages) 10 *Pieris rapae crucivora* and 10 *P. melete* 2nd instar larvae were put at random on the leaves of one and same cabbage and 10 females of *A. glomeratus* reared from larvae of *P. rapae crucivora* were introduced and left for 48 hr. The results obtained indicated that the average percentage of parasitization on *P. rapae* was 94 per cent and on *P. melete* 30.4 per cent. *A. glomeratus* is also reared on *Pieris rapae cruvivora* Boisduval by Tagawa (1984) in Japan and Sufty and Fuhrer (1981) on *Cydia pomonella* Durch.

A. angaleti was not studied with respect to host specificity except the record of some Lepidopterans like Litchi fruit borer *Conomorpha cramerella* and potato cutworms *A. ypsilon, A. segetum* and *Ecotomyelosis creatoninae* reported by Thompson (1951). In *A. angaleti* the host preference was much higher towards *A. ypsilon* than that of *A. mylitta* however, the parasitoid showed negative response to *S. litura, B. mori* and *P. brassicae.*

The host preferential studies will be helpful in understanding the alternative hosts of the silkworms and their stage of attack and the seasonal availability, further, these aspects would be beneficial for designing control measures of the natural enemies of silkworms.

V. LIFE TABLES AND INTRINSIC RATE OF INCREASE

Introduction

Organisms capacity for increase in numbers, 'r_m' has great importance in population ecology. 'r_m' is the inherent capacity for population increase of stable age distribution under given physical condition when competition and other biotic mortality factors are absent (Deevey, 1947). The actual rate of increase, 'r' in the rate of increase that pertains under any prevailing condition, and is a real statistic for population assessment. For comparing the potentials of different pest species, or of a pest species and natural enemies of pests, the intrinsic rate is very useful. According to Howe (1953), the control of insect pests of economic importance is the most pressing problem of recent years and for this the essential object is to make the estimate of the rate of growth of the pests and their natural enemies. An individual living animal in a given environment and from population will have a characteristic rate of development, longevity and fecundity at different stages in its life. The values of these means are determined in part by the environment and inherent characteristics of the living animal itself. The inherent characteristics of the living animal is "innate capacity for increase". As far as life table and intrinsic rate is concerned Thompson (1924) for the first time developed a mathematical method but proved to be very much laborious. However, Lotka (1925) derived a function for the intrinsic rate of natural increase, 'r_m'. Initially it was applied for human population but, Birch (1948) applied it to the insect population for the first time. Perusal of literature showed that Morris and Miller (1954), Stark (1959), Le Raux *et al.* (1963), Waloff (1968), Mc Leod (1972), Bains and Shukla (1976), Bilapate and Pawar (1980), Nikam and Sathe (1983b), Sathe (1986b, 1988c), Bhattacharya *et al.* (1993), Narayanswamy *et al.* (1993), Chakraborty *et al.* (1996c) etc. attempted life table studies in different orders of insects. In the present study the life tables were constructed with the help of fecundity data and later the intrinsic rates of increase of population of parasitoids were calculated by using Birch's (1948) formula as elaborated by Howe (1953) and Watson (1964).

Birch (1948) gave the equation $\Sigma e7\text{-}r_m X l_x m_x = 1$ for the intrinsic rate of natural increase where

'e' is base of the natural logarithms,

'x' is the age of the individual in days

'l_x' is the number of individuals alive at age 'x' as a proportion of one and

'm_x' is the number of female offsprings produced per female in the age interval 'x'.

The sum of products l_x m_x is the net reproductive rate, 'Ro', is the rate of multiplication of the population in each generation measured in terms of females produced per generation.

The approximate value of cohort generation time 'T_c' was calculated as follows :

$$T_C = \frac{l_X m_X x}{l_X m_X}$$

For arbitrary value of innate capacity for increase 'r_c' was calculated from the following formula :

$$r_c = \frac{\log_e Ro}{T_c}$$

The value of r_m up to two decimal places was added

in the formula until two values of the equation were found which lies immediately above or below 1096.6.

On the vertical axis arbitrary 'r_m' were taken while on the horizontal axis two values of

$$\Sigma e7\text{-}r_m x l_x m_x = 1$$

were plotted. The two points were jointed to give a line which interested a vertical line drawn from the desired value of 'r_m' accurate to 3 decimal places. The precise generation time 'T' was then calculated from the formula:

$$T = \frac{\log_e Ro}{r_m}$$

The finite rate of increase (λ) was calculated as

$$\lambda = e\ r_m$$

Material and Methods

Culture methods of parasitoids *Exorista bombycis* Louis *Apanteles glomeratus* Linn. *Apanteles angaleti* Muesebeck and their hosts namely *Bombyx mori* Linn. (Pm × NB_4D_2) and *Antheraea mylitta* Drury are given in the chapters 1-5 and 7-II. The parasitoids were exposed into rearing cages with constant number of their hosts and daily hosts were changed and the hosts used earlier were transferred separately in rearing cages type 1 (Fig. 19a). This was carried out till the death of female parasitoids. Thus, observations were made on the progeny production of mated females. Later, sex-ratio was counted. After collecting fecundity data, life tables were constructed according to Birch (1948) and intrinsic rate of increase was calculated in the species.

Results

E. bombycis

The number of progeny production averaged 335.5 (range, 266-434), adults longevity of ovipositing females ranged from 14 to 22 days (average 17.9 days), and sex-ratio, male : female, ranged from 1: 1.03 to 1:1.25 (average, 1:1.07). Average length of immature forms was 32 days. The maximum mean progeny production per day, 'm_x' was 20.4 on the 8th day and reproduction stopped on 24th day in a single generation, the intrinsic rate of increase in the species was 0.139 (Fig. 84) per female per day and population multiplied to 171.42 times in means generation time, 'T' of 37.008 days. The statistical results are shown in Tables 20, 21, 22, 23 and 24.

$$T_c = \frac{l_x m_x x}{l_x m_x} = \frac{6767.53}{171.42} = 39.479$$

where T_c is arbitrary T

$$= \frac{\log_e Ro}{T_c} = \frac{\log_e 171.42}{39.479}$$

where 'r_c' is arbitrary 'r_m'

$$T_c = 0.130$$

Now, arbitrary 'r_m's are 0.11 and 0.15

$\therefore$ 'r_m' = 0.139 (Fig. 84)

where λ is the finite rate of natural increase

$$T = \frac{\log_e 171.42}{0.139} = 37.008$$

T = 37.008 days.

A. glomeratus

Longevity of ovipositing females averaged 5.7 days (range 5 to 6 days). The number of progeny production averaged 166.6 (range, 152-180) adults with an average sex ratio, (male : female) 1:1.07, (range, 1:1.01-1:1.16). Average length of immature forms was 13 days and the maximum mean progeny production per day 'm_x' was 42.5 on the 2nd day and reproduction stopped on 7th day, in a single generation. The intrinsic rate of increase in the species was 0.292 (Fig. 85) per female per day and population multiplied to 86.2 times in mean generation time, 'T' of 15.262 days. Results are represented in Tables 25, 26, 27, 28 and 29.

$$T_c = \frac{l_x m_x x}{l_x m_x} = \frac{1340.5}{86.2} = 15.555$$

where T_c is arbitrary T

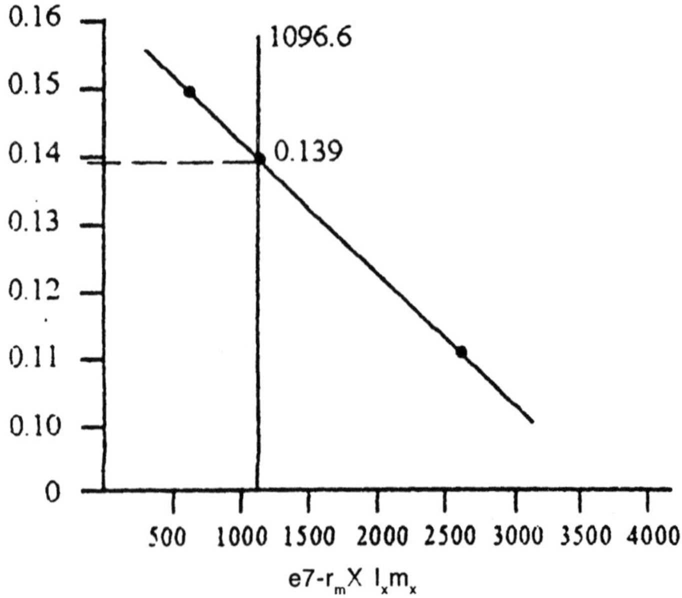

Fig. 84 : Deter..ination of intrinsic rate of increase in *E. bombycis*

Table 20 : Daily Female Production by Mated Females of *E. bombycis*

Days→ / Replicate↓	1	2	3	4	5	6	7	8	9	10	11	12	13	14	15	16	17	18	19	20	21	22	23	24	Total
A	4	7	8	10	11	13	15	18	16	14	9	9	8	4	3	D	–	–	–	–	–	–	–	–	149
B	3	5	7	9	12	20	26	18	17	19	18	15	4	2	1	1	D	–	–	–	–	–	–	–	177
C	3	7	9	11	16	21	18	20	9	8	6	4	3	3	2	1	2	2	1	D	–	–	–	–	146
D	7	10	10	12	14	15	12	27	11	10	9	7	6	6	5	4	3	1	3	1	1	1	1	D	175
E	6	11	14	15	18	27	28	30	19	17	10	9	5	3	2	1	1	1	2	2	1	1	D	–	223
F	2	7	10	14	16	18	14	18	16	13	10	9	7	7	5	4	3	3	1	1	D	–	–	–	178
G	4	9	8	13	13	15	15	24	10	8	5	5	4	1	1	D	–	–	–	–	–	–	–	–	135
H	10	7	16	19	26	17	28	18	10	8	9	6	5	4	4	3	2	2	1	D	–	–	–	–	195
I	8	9	10	10	16	17	25	13	11	12	9	6	3	2	2	3	2	1	D	–	–	–	–	–	159
J	5	6	17	18	21	24	19	18	15	13	10	8	6	4	3	4	2	3	1	1	1	D	–	–	199
Average	5.2	7.8	10.9	13.1	16.3	18.7	20	20.4	13.4	12.2	9.5	7.8	5.1	3.6	2.7	2.1	1.5	1.3	0.9	0.5	0.3	0.2	0.1	0	173.6

$$r_c = \frac{\log_e Ro}{T_c} = \frac{86.2}{15.555} = 0.286$$

where 'r_c' is arbitrary 'r_m'

T_c = 0.286

Now, arbitrary 'r_m's are 0.26 and 0.30

∴ 'r_m' = 0.292 (Fig. 85)

where λ is the finite rate of natural increase

$$T = \frac{\log_e 86.2}{0.292} = 15.262$$

T = 15.262 days.

Table 21 : Longevity, Oviposition : Sex Ratio and Fecundity of Mated Females of *E. bombycis*

Female No.	No. of larvae exposed	Longe-vity (days)	Oviposition days	Parasitoid progeny			Sex-ratio M:F
				Male	Female	Total	
A	400	16	15	140	149	289	1:1.06
B	400	17	16	171	177	348	1:1.03
C	400	23	22	140	146	286	1:1.04
D	400	17	15	139	175	314	1:1.25
E	400	19	18	211	223	434	1:1.05
F	400	20	19	169	178	347	1:1.05
G	400	21	20	131	135	266	1:1.03
H	400	20	19	182	195	377	1:1.07
I	400	22	21	149	159	308	1:1.06
J	400	16	14	187	199	386	1:1.06
Average =		19.1	17.9	161.9	173.6	335.5	1:1.07

A. angaleti

The duration for immature stages was 13 days, oviposition days averaged 5.6 days (range, 5-6 days). The progeny production averaged 154.4 individuals (range, 142-170 adults) with an average sex ratio, male : female, 1 : 1.02 (range 1:0.94 to 1:1.11) offsprings. The m_x i.e. maximum mean progeny production per day was 42.1 on 2nd day and production stopped on the 7th day. The innate capacity for increase was found to be 0.285 (Fig. 86) per female per day and population multiplied to 78.2 times in a generation time of 15.295 days. Results are tabulated in Table 30, 31, 32, 33 and 34.

Table 22 : Life Table Statistics of *E. bombycis*

Pivotal age (days) x	Proportional life at age x l_x	Number of female progeny/ famale, m_x	$l_x m_x$	$l_x m_x X$
1 to 32 days immature stages				
33	1	5.2	5.2	171.6
34	1	7.8	7.8	265.2
35	1	10.9	10.9	381.5
36	1	13.1	13.1	471.6
37	1	16.3	16.3	603.1
38	1	18.7	18.7	710.6
39	1	20.0	20.0	780.0
40	1	20.4	20.4	816.0
41	1	13.4	13.4	549.4
42	1	12.2	12.2	512.4
43	1	09.5	09.5	408.5
44	1	07.8	07.8	343.2
45	1	05.1	05.1	229.5
46	1	03.6	03.6	165.6
47	1	02.7	02.7	126.9
48	0.9	02.1	01.89	090.7
49	0.7	01.5	01.05	051.4
50	0.7	01.3	00.91	045.5
51	0.6	00.9	00.54	027.5
52.	0.4	00.5	00.20	010.4
53	0.3	00.3	00.09	004.77
54	0.2	00.2	00.40	002.16
55	0.0	00.1	00.00	000.00
			171.43	Ro = Σ6767.53

Table 23 : Provisional r_m (.11) for *E. bombycis* and Related Values of $e7\text{-}r_m x l_x m_x$

x	rm_x	$7\text{-}r_m x$	$e7\text{-}r_m x$	$e7\text{-}r_m x l_x m_x$
33	3.63	3.37	29.078	151.205
34	3.74	3.26	26.049	203.182
35	3.85	3.15	23.336	254.362
36	3.96	3.04	20.905	273.855
37	4.07	2.93	18.727	305.250
38	4.18	2.82	16.776	313.711
39	4.29	2.71	15.029	300.580
40	4.40	2.60	13.463	274.645
41	4.51	2.49	12.061	161.617
42	4.62	2.38	10.804	130.808
43	4.73	2.27	09.679	091.950
44	4.84	2.16	08.671	067.633
45	4.95	2.05	07.767	039.611
46	5.06	1.94	06.958	025.048
47	5.17	1.83	06.049	016.332
48	5.28	1.72	05.584	010.553
49	5.39	1.61	05.002	005.252
50	5.50	1.50	04.481	004.077
51	5.61	1.39	04.014	002.167
52	5.72	1.28	03.596	000.719
53	5.83	1.17	03.221	000.289
54	5.94	1.06	02.886	000.115
55	6.05	0.95	02.585	000.000
				Ro = Σ 2632.961

$$T_c = \frac{l_x m_x x}{l_x m_x} = \frac{1212.3}{78.2} = 15.502$$

where T_c is arbitrary T

$$r_c = \frac{\log_e Ro}{T_c} = \frac{78.2}{15.502} = 0.281$$

where 'r_c' is arbitrary 'r_m'

$T_c = 0.281$

Now, arbitrary 'r_m's are 0.26 and 0.30

$\therefore$ 'r_m' = 0.285 (Fig. 86)

where λ is the finite rate of natural increase

$$T = \frac{\log_e 78.2}{0.285} = 15.295$$

T = 15.295 days.

Table 24 : Provisional r_m (.15) for *E. bombycis* and Related Values of $e7\text{-}r_m x l_x m_x$

x	rm_x	$7\text{-}r_m x$	$e7\text{-}r_m x$	$e7\text{-}r_m x l_x m_x$
33	4.95	2.05	7.767	40.388
34	5.10	1.90	6.685	52.149
35	5.25	1.75	5.754	62.725
36	5.40	1.6	4.953	64.884
37	5.55	1.45	4.263	69.488
38	5.7	1.3	3.669	68.615
39	5.85	1.15	3.158	63.163
40	6.00	1.00	2.718	55.452
41	6.15	0.85	2.339	31.351
42	6.30	0.70	2.013	24.558
43	6.45	0.55	1.733	16.465
44	6.60	0.40	1.491	11.636
45	6.75	0.25	1.284	6.458
46	6.9	0.10	1.105	3.978
47	7.05	–0.05	1.051	–2.838
48	7.2	–0.02	1.221	–2.308
49	7.35	–0.35	1.419	–1.490
50	7.5	–0.50	1.648	–1.500
51	7.65	–0.65	1.915	–1.034
52	7.8	–0.80	2.225	–0.445
53	7.95	–0.95	2.585	–0.232
54	8.1	–1.1	3.004	–0.120
55	8.25	–1.25	3.490	0.000
				Ro = Σ 561.343

Table 25 : Daily Female Production by Mated Females of *A. glomeratus*

Days → Replicate ↓	1	2	3	4	5	6	7	Total
A	10	44	15	11	8	2	D	90
B	12	41	21	10	7	1	D	92
C	9	36	17	8	7	1	D	78
D	13	39	15	10	6	D	–	83
E	11	42	13	8	8	1	D	83
F	11	47	14	11	6	2	D	91
G	10	40	17	12	7	D	–	86
H	8	48	17	10	5	2	D	90
I	9	47	15	9	4	D	–	84
J	12	41	18	10	6	1	D	88
Average	10.5	42.5	16.2	9.9	6.4	1.0		86.5

Table 26 : Longevity, Oviposition, Sex Ratio and Fecundity of Mated Females of *A. glomeratus*

Female No.	No. of larvae exposed	Longe-vity (days)	Oviposition days	Number of parasitoid progeny			Sex-ratio M : F
				Male	Female	Total	
A	100	7	6	87	90	177	1:1.03
B	100	8	6	88	92	180	1:1.04
C	100	7	6	74	78	152	1:1.05
D	100	6	5	75	83	158	1:1.10
E	100	7	6	71	83	154	1:1.16
F	100	7	6	83	91	174	1:1.09
G	100	6	5	78	86	164	1:1.10
H	100	7	6	78	90	168	1:1.15
I	100	6	5	83	84	167	1:1.01
J	100	7	6	84	88	172	1:1.04
Average =		6.8	5.7	80.1	86.5	166.6	1:1.07

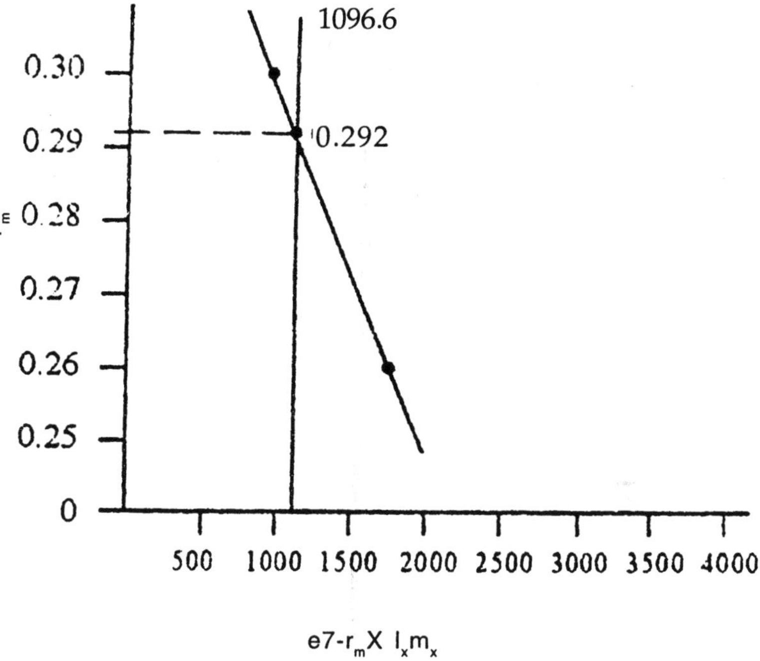

Fig. 85 : Determination of intrinsic rate of increase in *A. glomeratus*

Table 27 : Life Table Statistics of *A. glomeratus*

Pivotal age (days) x	Proportional life at age x l_x	Number of female progeny/ female, m_x	$l_x m_x$	$l_x m_x X$
1 to 13 days immature period				
14	1	10.5	10.5	147.00
15	1	42.5	42.5	637.50
16	1	16.2	16.2	259.20
17	1	09.9	09.9	168.30
18	1	06.4	06.4	115.20
19	0.7	01.0	00.7	013.30
20	0	00.0	00.0	000.00
			Average = 86.2	Ro = Σ1340.5

Table 28 : Provisional r_m (0.26) for *A. glomeratus* and Related Values of $e7\text{-}r_m x l_x m_x$

x	$r_m x$	$7\text{-}r_m x$	$e7\text{-}r_m x$	$e7\text{-}r_m x l_x m_x$
14	3.64	3.56	28.789	302.284
15	3.90	3.10	22.197	943.372
16	4.16	2.84	17.115	277.263
17	4.42	2.58	13.197	130.650
18	4.68	2.32	10.175	065.120
19	4.94	2.06	07.845	005.491
20	5.20	1.80	09.049	000.000
				Ro = Σ 1724.18

Table 29 : Provisional r_m (0.30) for *A. glomeratus* and Related Values of $e7\text{-}r_m x l_x m_x$

x	$r_m x$	$7\text{-}r_m x$	$e7\text{-}r_m x$	$e7\text{-}r_m x l_x m_x$
14	4.2	2.8	16.444	172.662
15	4.5	2.5	12.182	517.735
16	4.8	2.2	09.025	146.205
17	5.1	1.9	06.685	066.181
18	5.4	1.6	04.953	031.699
19	5.7	1.3	03.669	002.258
20	6.0	1.0	02.718	000.000
				Ro = Σ 936.74

Discussion

Narayanswamy *et al.* (1993) studied the life table of uzi fly *E. bombycis* as per Harcourt (1969) for rainy, winer and summer seasons. The mean life table over 18 generations revealed mortality of 13.32, 14.6, 5.85, 20.96 and 9.54 per cent at egg, maggot, pre-pupal, pupal and adult stages, respectively. Infertility was found to be the key mortality factor at egg stage, while the failure of emergence of maggots from superparasitized host, diseased host and host cocoon contributed more towards maggot mortality. The

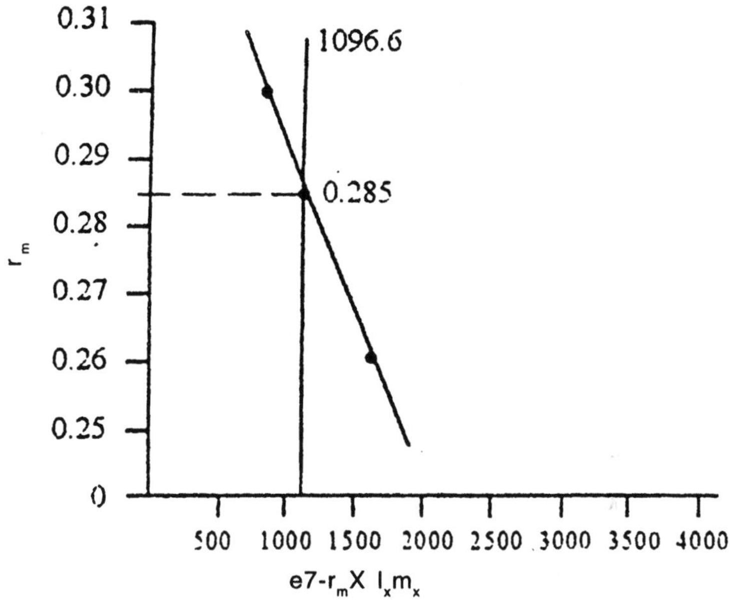

Fig. 86 : Determination of intrinsic rate of increase in *A. angaleti*

hyperparasitoids *Nesolynx thymus* (Girault) and *Trichopria* sp. have been recorded to be key mortality factors affecting population at pupal stage. The adults with deformed wings and legs and also heavily sclerotized adults that were unable to reproduce, caused reduction of uzi fly population. The egg mortality was lowest followed by that of maggot, pre-pupal, pupal and adult. But all the curves were in accordance with the type-I curve of Slobodkin (1962) indicating that the mortality occurred in the later stages of development cycle of uzi fly. The mortality factors were almost similar in all seasons. However, perusal of the overall death rate curve showed that the mortality rate was highest during pupal stage followed by that in maggot, egg, adult and pre-pupal stage.

In the present study life tables of *E. bombycis* were constructed as per Birch (1948) under laboratory conditions (25±1°C, R.H. 75±3, 18:6 hr L:D) wherein instrinsic rate of natural increase was 0.139 per female per day and population multiplied to 171.42 times in mean generation time of 37.008 days. The maximum mean progeny production per day, m_x' was 20.4 on the 8th day and reproduction stopped on 24th day.

Table 30 : Daily Female Production by Mated Females of *A. angaleti*

Days → Replicate ↓	1	2	3	4	5	6	7	Total
A	8	45	12	9	6	2	D	82
B	11	42	18	8	7	1	D	87
C	15	38	13	10	5	1	D	82
D	7	39	11	8	6	D	–	71
E	8	41	9	9	7	1	D	75
F	11	42	10	8	5	D	–	76
G	10	47	8	7	6	2	D	80
H	6	42	8	6	7	1	D	70
I	9	44	13	9	2	D	–	77
J	9	41	16	11	6	D	–	83
Average	9.4	42.1	11.8	8.5	5.7	0.8		78.3

Table 31 : Longevity, Oviposition, Sex Ratio and Fecundity of Mated Females of *A. angaleti*

Female No.	No. of larvae exposed	Longevity (days)	Oviposition days	Number of parasitoid progeny			Sex-ratio M : F
				Male	Female	Total	
A	100	7	6	80	82	162	1:1.02
B	100	7	6	83	87	170	1:1.04
C	100	7	6	79	82	161	1:1.03
D	100	6	5	71	71	142	1:1.00
E	100	7	6	70	75	145	1:1.07
F	100	6	5	68	76	144	1:1.11
G	100	7	6	77	80	157	1:1.03
H	100	7	6	74	70	144	1:0.94
I	100	6	5	78	77	155	1:0.98
J	100	6	5	81	83	164	1:1.02
Average		6.6	5.6	76.1	78.3	154.4	1:1.02

Table 32 : Life Table Statistics of *A. angaleti*

Pivotal age (days) x	Proportional life at age x l_x	Number of female progeny/ famale, m_x	$l_x m_x$	$l_x m_x X$
1 to 13 days immature days				
14	1	9.4	9.4	131.6
15	1	42.1	42.1	631.5
16	1	11.8	11.8	188.8
17	1	8.5	8.5	144.5
18	1	5.7	5.7	102.6
19	0.8	0.88	0.70	13.3
20	0.0	00	00	00
				Ro = Σ 1212.3

Recently, Chakraborty *et al.* (1996c) constructed the life table of *E. bombycis* through 3 generations using 2nd stage larvae of *B. mori* as a host for depositing eggs within the different biotic characters like mortality at every stage of its life cycle, pupal period, adult longevity, fecundity, fertility and sex ratio in relation to the abiotic factors (temperature and relative humidity) prevailing in the rearing room. Generation survivability was recorded to be varied from 4.44 per cent to 7.04 per cent. Maximum mortality at egg, larval, maggot and pupal stages was 66.02 per cent, 46.94 per cent, 7.69 per cent and 20.83 per cent respectively. Fertility and sex-ratio were found to be varied from 28.57 per cent to 33.33 per cent and 1 : 0.77 to 1 : 1.1 respectively. Correlation of coefficient between the biotic and abiotic factors was also studied by these workers. Their results indicated that the biotic factors also attributed to high mortalities in eggs and larvae. In the present study life tables were studied by providing 10 day old larvae under laboratory conditions.

Table 33 : Provisional r_m (0.26) for *A. angaleti* and Related Values of $e7\text{-}r_m x l_x m_x$

x	$r_m x$	$7\text{-}r_m x$	$e7\text{-}r_m x$	$e7\text{-}r_m x l_x m_x$
14	3.64	3.36	28.789	270.616
15	3.90	3.16	22.197	934.493
16	4.16	2.84	17.115	201.957
17	4.42	2.58	13.197	112.174
18	4.68	2.32	10.175	57.997
19	4.94	2.06	7.845	5.491
20	5.2	1.80	6.049	0.00
				Ro = Σ1582.674

As the intrinsic rate of natural increase (r_m) can be used for gaining in sight into the population dynamics of a species, several workers attempted such studies in different orders of the insects. Richards and Waloff (1961) on *Phytodecta olivacea* (Coleoptera), Waloff (1968) on *Sitonaregan steinansis* Herbst (Coleoptera) and *Arytaina genistae* (Hemiptera), McLeod (1972) on *Neodiprion swainei* Midd. (Hymenoptera), Moris and Miller (1954) on *Choristoneura fumiferana* (Lepidoptera), Stark (1959) on *Recurvaria starki* (Lepidoptera) and Le Roux *et al.* (1963) on *Spilonota ocellava* (Lepidoptera).

Table 34 : Provisional r_m (0.30) for *A. angaleti* and Related Values of $e7\text{-}r_m x l_x m_x$

x	$r_m x$	$7\text{-}r_m x$	$e7\text{-}r_m x$	$e7\text{-}r_m x l_x m_x$
14	4.2	2.8	16.444	154.573
15	4.5	2.5	12.182	512.862
16	4.8	2.2	9.025	106.495
17	5.1	1.9	6.685	56.822
18	5.4	1.6	4.953	28.232
19	5.7	1.3	3.669	2.568
20	6.0	1.0	2.718	0.00
				Ro = Σ 861.552

In *Cotessia flavipes* (Cameron) (Hymenoptera, Braconidae) the intake capacity of increase per female per day was 0.176 and population multiplied to 30.72 times in mean generation time of 19.45 days (Nikam and Sathe, 1983b). Likely, in *C. diurnii* C. & R., a parasitoid of *Exelastis atomosa* Wals. the intrinsic rate of increase was 0.158 and population multiplied 25.99 times in mean generation time (T) 20.61days. The maximum mean progeny production per day, m_x' was 9.70 on the 3rd day (Sathe, 1986b). In the present braconid form (*A. glomeratus*) the intrinsic rate• of increase was 0.292 per female per day and population multiplied to 86.2 times in mean generation time, of 15.262 days. The maximum mean progeny production per day m_x was 42.5 on the 2nd day and reproduction stopped on 7th day.

Sathe (1988c), studied the intrinsic rate of increase and interspecific relationships between *Cotesia orientalis* C. & N., *C. diurnii* R. & N. and *Diadegma trichoptilus* Cameron (Hymenoptera), the larval parasitoids of *E. atomosa,* wherein the intrinsic rates were 0.188, 0.158 and 0.149 and population multiplied to 41.93, 25.99 and 25.63 times in mean generation time of 19.87, 20.61 and 21.77 days respectively while, in *A. angaleti* the intrinsic rate of increase was 0.285 and the population multiplied to 78.2 in mean generation time of 15.295 days. The m_x i.e. maximum mean progeny production per day was 42.1 on 2nd day and production stopped on the 7th day.

Recently, Jyothi and Veeranna (1993) studied a pupal parasitoid *Dirhinus anthracia* (Chalcidae, Hymenoptera) with respect of life tables wherein they noted that the rate of reproduction per generation was found to be 125.029 while, appropriate rate of natural increase (r_c) was 0.1021. The intrinsic rate of natral increase (r_m) was found to be 0.1132 and respective generation time was 42.63 days. The finite rate of increase was 1.119 indicating that the population of *D. anthracia* multiplies 1.119 times per female per day. Similarly, Sathe and Ingawale (1993b) studies the braconid wasp, *Apanteles jayanagarensis* Bhatnagar, a parasitoid of Bihar hairy caterpillar, *Spilosoma obliqua* (Wlk.) and reported that the number of progeny produced ranged from 108 to 134 individuals ($\bar{x}$ 119.6). The male and female offsprings ranged from 1 : 0.95 to 1 : 2.051 ($\bar{x}$ 1 : 1.349). The first adult mortality was on the 5th day. The average length of immature stages of parasitoids was 16 days. The maximum mean progeny production per day (m_x) was 18.8 intrinsic rate (r_m) per female per day was 0.225 and population multiplied 66.71 times in mean generation time of 18.66 days. Sathe (1990c) also studied one more parasitoid, *Parenion bhairvi* S. & I. of *S. obliqua* wherein longevity of ovipositing females averaged 7.7 days, progeny production averaged 70.4 individuals with m : f offspring, 1 : 1.24. The average period of immature stages was 19 days, the maximum mean female progeny production per day (mx) was 12.7 on the 2nd day. The innate capacity for increase in this parasitoid was 0.177 per female per day and population multiplied 57.89 times in mean generation time of 20.53 days.

The present study would be helpful for understanding the life table statistics and adopting the control measures of parasitoids studied here.

Chapter 8
SUMMARY

Sericulture is increasing at a very rapid rate in India since it is related to employment, commerce and economic development of country and is a potential source of foreign exchange. Owing to the fast spread of sericulture to the most of states with heterogenous agroclimatic conditions, various mulberry cultivation methods and silkworm rearing methods have been evolved and being practised in India. However, the expected quality and yield of cocoon is not achieved so far because of the damage caused by certain pests and diseases to silkworms. A very large number of natual enemies are recorded from insect and non-insect groups. However, uziflies (Diptera) and Hymenopterous wasps cause considerable damage to silkworms. Hence, information is given on *Exorista bombycis* (Diptera), *Apanteles angaleti* and *Apanteles glomeratus* (Hymenoptera) *E. bombycis* and *A. glomeratus* are larval parasitoids of mulberry silkworm (*B. mori*) while; *A. angaleti* of tasar silkworm (*A. mylitta*).

The first chapter is devoted to general introduction which narrates national and international status of sericulture and its development. The second chapter deals with Moriculture. Mulberry cultivation, fertilizers, pests, diseases and their control, are the important aspects of this chapter. The third and the fourth chapters provide information regarding disinfectants and seed production respectively. The fifth chapter is devoted to silkworm rearing in which mulberry and tasar silkworm rearing techniques are described. In addtion, information is given on silkworm pests, diseases and their control. The sixth chapter refers to silkworm natural enemy survey and survelliance while, the seventh includes aspects of silkworm parasitoids under which life history, rearing, host age selection, host specificiy and intrinsic rates of natural increase have been reported.

BIBLIOGRAPHY

Anonymous, 1889. Notes on Indian economic entomology: I - *Trycolyga Bombycis*. *Indian Museum Notes*, **1**, 77-81, 83-88.

Anonymous, 1975. *Text book of Tropical Sericulture*, Japan Overseas Co-operation Volunteers, Tokyo, Japan, 540-547.

Anonymous, 1980. The winged menance. *Indian Silk*, **14**, 5-6.

Anonymous, 1981. Uzi fly control (In Karnataka). *Reshme Krishi*, **3**, 11.

Anonymous, 1982. Survey and surveillance project on uzi fly. *CSR&TI*, Mysore, 8.

Anonymous, 1985. CSB, *Indian Silk*, **23**, 33.

Anonymous, 1989. Survey on the incidence of uzi fly at research level *Annu*silkworm *CSR & TI*, Mysore, 58.

Anonymous, 1993. Bibliography on uzi fly research. *CSR & TI*, Mysore, 1-58.

Anonymous, 1998. MITCON, 1-104.

Arthur, A.P., 1966. Associative learning in *Itoplectis conguisitor* (Say.) (Hymenoptera, Ichneumonidae). *Can. Ent.*, **98**, 213-223.

Ayuthaya, N., 1972. The host preference of the tachinid fly to different stages and varieties of silkworm. *Bull Thai Sericult. Res. Train. Centre*, **2**, 67-69.

Ayuzawa, C., Sekido, D., Yamakawa, K., Sakurai, U., Kurato, W., Yaginuma Y. and Takoro (1972). *Handbook of silkworm rearing*. Fuji Pub. Co., Ltd., Tokyo, Japan, 319.

Azuma, Kan-Ichi and Hideo Kitano 1971. Experimental studies on the parasitism of *Apanteles glomeratus* Linn. on the larvae of *Pieris melete* Menetries. *Kontyu,* **39,** 394-399.

Barman, A.C., Pasha, K. and Nạhar, A., 1990. A study on the control of uzifly *Tricholyga bombycis* (Beck) in silkworm, *Bombyx mori* L. *Bull. Seric. Res.,* **1,** 65-69.

Bartlett, B.R., 1953. A tactile ovipositional stimulus to culture *Macrocentrus ancylivorus* on an unnatural host. *J. Econ. Entomol.,* **46,** 525.

Beeson, C.F.C. and Chatterjee, S.N., 1935. The biology of *Tachinidae* (Diptera). *Indian Forest Records,* **1,** 184.

Benchamin, K.V. and Jolly, M.S., 1987. Employment and income generation in the rural areas through Sericulture. *Indian Silk,* **25,** 9-13.

Bharali, N., 1968. Some observations of fly pest infestation of muga silkworm *Antheraea assama. Indian J. Seric.,* **7,** 42-44.

Bhattacharya, S.S., Chakraborty, N. and Sahakundu, A.K., 1993. Life table of the uzifly, *Exorista sorbillans* (Wiedemann) (Diptera : Tachinidae), a parasitoid of the mulberry silkworm, *Bombyx mori. Sericologia,* **33,** 65-79.

Bhide, S., 1994. Sericulture in Andhra Pradesh by 2001, Souvenir, International Conference on Sericulture : *Global Silk Scenario 2001,* 125-129.

Bilapate, G.G. and Pawar, V.M., 1980. Life fecundity tables for *Heliothis armigera* Hubner (Lepidoptera, Noctuidae) on sorghum earhead. *Proc. Indian Acad. Sci.* (Ani. Sci.), **89,** 69-73.

Birch, L.C., 1948. The intrinsic rate of natural increase in an insect population. *J. Anim. Ecol.,* **17,** 15-26.

Broodryk, S.W., 1969. The biology of *Chelonus (Microchelonus) curvimaculatus* Cameron (Hymenoptera : Braconidae). *J. ent. Soc. Sth. Afr.,* **32,** 169-189.

Carlson, R.W., 1972. Suppression of the name, *Campoletis perdistinctus* (Hymenoptera, Ichneumonidae) and the identity of species to which the name has been applied. *Ent. News,* **83,** 875-82.

Chakraborty, N., Bhattacharya, S.S., Das, N.K., and T. Pawankumar, 1996a. Studies on incidence of uzi fly (*Exorista bombycis* Louis) infestation on silkworm (*Bombyx mori* L.) and assessment of cocoon crop loss in Murshidabad district (West Bengal, India). *Entomon,* **21,** 29-35.

Chakraborty, N., Bhattacharya, S.S., Das, N.K., Sen, S.K., and T. Pawankumar. 1996b. Survey and the seasonal incidence of uzi fly *Exorista bombycis* (Louis) infestation on the silkworm, *Bombyx mori* L. and assessment of cocorn loss in Birbhum district West Bengal. *Sericologia,* **36,** 703-709.

Chakraborty, N., Bhattacharya, S.S., Sahakundu, A.K. and S.K. Sen, 1996c. Life table of uzi fly *Exorista bombycis* Louis (Diptera : Tachinidae) parasitizing 2nd stage larvae of silkworm, *Bombyx mori. Entomon,* **21,** 243-250.

Chattarjee, S. and Swarup, P. 1961. *Apanteles belippae* Rohwer. (Hymenoptera, Braconidae), a new natural enemy of silkworm, *Bombyx mori* L. *Indian J. Ent.,* **23,** 157-158.

Chaudhury, S.N., 1984. *Mulberry Silk Industry*. Greham Dibrugarh, Assam, India, 270.

Coppel, H.C. and Mertins, J.W., 1977. *Biological insect pest superssion. Adv. Series. Agri. Sci.*, Springer-Verlag Berlin Heidelberg, New York. 1-301.

Cotes, E.C., 1889. Further notes on Insect pests 14. Short notes on miscellaneous insects pests, Diptera. *Indian Museum., Notes,* **1,** 63-64.

Cotes, E.C., 1896. Miscellaneous Notes, *Indian Museum, Notes,* **3,** 16-33.

Crosskey, R.W., 1977. Tachinidae, In: *A Catalogue the Diptera of the Oriental region*. (M.D. De lfinado and D.E. Hardy Eds.), The University Press of Hawaii, Honoulu, **3,** 586-697.

Dandin, S.B. and Ramesh, S.R., 1987. *A kalpa vruksha* called mulberry. *Indian Silk,* **26,** 49-53.

Das Gupta, Abhijit, 1994. Sericulture in Karnataka by 2001. Souvenir, International Conference on Sericulture *Global Silk Scenario 2001,* 121-124.

Das, P.K., Ghosh, A., Choudhury, P.C., Katiyar, R.S. and Sengupta, K. 1992. Responses of irrigated mulberry to Azotobacter and Azospirillum biofertilizer under graded levels of nitrogen in biofertilizer technology transfer (L.V. Gangawane, Edn.), Associated Publishing Company, New Delhi, 71.

Das, P.K., Choudhary, P.C., Ghosh, A., Katiyar, R.S., Mathur, V.B. and Datta R.K., 1993. Use of *Azotobacter* biofertilizer in mulberry cultivation. *Indian Silk,* **31,** 43.

Datta, R.K., 1995. Sericulture biotechnology, *Proc. Nat. Acad. Sci. India,* (B), II (suppl.) 201-216.

Datta, R.K., Biswas S.N., Haques, S. and Mishra P.N., 1977. Population dynamics of the silkworm pest *Trycholyga bombycis* Beck. (uzi fly). abs. *Semi. Insect and Envl.,* Delhi Univ., Delhi.

Datta, R.K. and Mukherjee, P.K., 1978. Life history of *Tricholyga bombycis* (Diptera, Tachinidae), a parasite of *Bombyx mori* L. (Lepidoptera, Bombycidae). *Ann. ent. Soc. Am,* **71,** 767-770.

Deevey, E.S. Jr., 1947. Life tables for natural populations of animals, *Quart. Rev. Bid.,* **22,** 283-314.

Dethier, V.G., 1961. Behavioural aspects of protein inquestion by the blowfly, *Phormia reginae* Meigen. *Biol. Bull.mar. biol. Lab.Woods Hole.,* **121,** 456-470.

Devaiah, M.C., Govindan, R. and Narayanaswamy, K.C., 1993. Life tables of uzi fly *Exorista bombycis* In: *Recent Advances in Uzi Fly Research* Pro. Nat. Semi. Uzi fly & its control, 1992. Bangalore. 1-12.

Doutt, R.L., 1959. The biology of parasitic Hymenoptera. *A. Rev. Ent.,* **4,** 161-182.

Drooz, A.T. and Fedde, V.H., 1972. Discriminate host selection by *Monodontomerus dentipes. Environ. Entomol.,* **1,** 522-523.

Erulen, J.M., 1966. The role of time and energy in food preference. *Amer. Nat* **100,** 611-617.

FAO, 1987. *Mannuals on sericulture,* **2,** 1-9.

Fisher, R.C., 1959. Life history and ecology of *Horogenes chrysostictes* Gmelin, a parasitoid of *Epthestia sericurium* Scott. (Lepidoptera : Phycitidae). *Can. J. Zool.,* **37,** 429-446.

Fisher, R.C., 1961. A study on insect multiparasitism. I. host selection and oviposition, *J. Exp. Biol.,* **38,** 267-275.

Fulton, B.B., 1940. The horn worm parasite, *Apanteles congregatus* (Say) and the hyperparasite, *Hypoteromalus tabacum* (Fitch.). *Ann, ent. soc. Am.*, **33,** 361-362.

Gangawar, S.K. and Thangavelu, K., 1989. Uzi fly infestation on young age silkworms of *Bombyx mori* L. in Tamil Nadu, India. *Entomon,* **14,** 361-362.

Gangrade, G.A., 1964. On the biology of *Campoletis perdistinctus* (Hymenoptera, Ichneumonidae) in Madhya Pradesh. *Ann. ent. Soc. Am.,* **57,** 570-574.

Gayathri Devi, K.G., 1994, Karnataka: A status report on women in sericulture. *Indian Silk,* **32,** 31-38.

Ghosh, C.C., 1933, The Silk Industry of Japan with notes on observations in the United States of America, England, France and Italy, Managers of Publication, New Delhi, 126.

Govindaiah, Sharma, D.D., Bajpai, A.K. and Datta, R.K., 1997, Mulbery cultivation in China. *Indian Silk,* **36,** 41-46.

Govindaraju, R., Saratchandra, B. and Krishnan, S., 1990. A field study of the influence of climatic factors on the infestation of Silkworm *Bombyx mori* L. by uzi fly *Exorista sorbillans* (Wiediemann) in different seasons. *Indian J. Seric.,* **29,** 284-288.

Gutierrez, A.P., 1970. Studies on host selection and host specificity of the aphid hyperparasite *Charips victrix* 5. Host selection. *Ann. ent. Soc. Am.* **63,** 1495-1498.

Hiromu, A., 1998. Global scenario of wild silks. *Indian silk,* **37,** 18-20.

Hopper, K.R. and King, E.G., 1984. Preference of *Microplitis croceipes* (Hymenoptera, Bracoonidae) for instars and species for *Heliothis* (Lepidoptera, Noctuidae). *Environ. Entomol,* **15,** 1145-1150.

Howe, R.W., 1953. The rapid determination of intrinsic rate of increase of an insect population. *Appl. Biol.*, **40,** 134-155.

Hu, C., 1982. A survey of the parasite of the small white butterfly, *Artogeia rapae* L. in China. *Acta Entomologia Sinica,* **16,** 287-294.

Isarangkul, N.A., Ayuthaya, L. and Sinchaisir, N., 1971. Notes on the parasitic potency of the tachinid fly, *Tachina sorbillans* Wied. to different stages of the silkworm. *Bull. Thai. Seric. Res. Train. Centre,* **1,** 98-101.

Isarangkul, N.A., Ayuthaya, L., Sinchaisir, N., and Aoki, K., 1972. On some observations of the tachinid fly. *Bull. Thai. Seric. Res. Train Centre,* **2,** 89-52.

Ishikawa, K., 1934. On the morphology of *Tricholyga sorbillans* Wied. *Zool. Mag. Tokyo,* **46,** 43-52.

Iwasa, Y., Suzuki, Y. and Matsuda, H., 1984. Theory of oviposition strategy of parasitoids I. Effect of mortality and limited egg number. *Theo. Pop. Biology,* **26,** 205-227.

Jackson, C.G., Neemann, E.G. and Patana, R. 1979. Parasitization of 6 Lepidopteran cotton pests by *Chelonus blackburni* (Hymenoptera, Braconidae). *Entomophaga,* **24,** 99-105.

Jones, D., Jones, G., van Steenwyk, R.A. and Hammock, B.D., 1982. Effect of the parasite *Copidosoma truncatellum* on development of its host, *Trichoplusia ni. Ann. ent. Soc. Am.,* **64,** 202-205.

Jolly, M.S. 1967. A brief report on wild sericigena in India with special reference to tasar culture world meet. *Silk Product, Mauiracia,* 1-6.

Jolly, M.S. 1998. Oak Tasar in India: Problems and prospects. *Indian Silk*, **37**, 35-37.

Jolly, M.S. and Chandrashekharaih, 1982. Field survey of uzi fly infestation in Karnataka. *Workshop on uzi fly control*. CSR & TI, Mysore, 1-6.

Jyothi, H.K. and Veeranna, G., 1993. Studies on the biology of *Dirhinus anthracia* (Hymenoptera, Chalcididae), a pupal parasitoid of *Exorista sorbillans* (Diptera, Tachinidae) and its life table. *Recent Advances in Uzi Fly Research*. Proceedings of National Seminar on uzi fly and its control. Eds. G.P. Channa Basavanna, G. Veeranna and S.B. Dandin, 143-153.

Kajita, H. and Drake, E.F., 1969. Biology of *Apanteles chilonis* and *Apanteles flavipes* (Hymenoptera, Braconidae), parasites of *Chilo suppressalis*. *Mushi*, **42**, 163-174.

Kasturibai, A.R., Mahadevappa, D., Nirmala, M.R. and Jyothi, H.K., 1986. Control of uzi fly by semiochemicals. *Curr. Sci.*, **55**, 1038-1040.

Kishi, Y., 1970. Difference in the sex ratio of the pine bark weevil parasite, *Dolichonitus* sp. (Hym., Ichneumonidae) emerging from different host species. *Appl. Ent. Zool.*, **5**, 126-132.

Kole, S. and Chaterjee, H. 1995. New endoparasite of *A. mylitta* D. *Indian Silk*, **34**, 33-34.

Krishna Rao, J.V., Singh, R.N. and Singh, C.M., 1998. Tribal development and wild silks. *Indian Silk*, **37**, 79-83.

Krishnaswami, S., 1978. New technology of silkworm rearing. *Bulletin, CSR & TI*, Mysore, **2**, 1-23.

Krishnaswami, S., Jolly, M.S., and Datta, R.K., 1964. A study on the fly pest infestation of the larvae and cocoon of *Bombyx mori* L. *Indian J. Seric.*, **3**, 7-12.

Krishnaswami, S., Narasimhanna, M.N., Suryanarayan, S.K. and Kamaraj, S., 1973. *Manual on sericulture,* **2** Silkworm Rearing. FAO. *Agri. Service Bull.*, **15, 131**.

Kumar Pradip, and Jolly, M.S., 1986a. Studies on the ovipositional preference of uzi fly *Tricholyga bombycis* Beck (Diptera, Tachinidae). *Indian J. Seric.*, **25**, 64-68.

Kumar Pradip, and Jolly, M.S., 1986b., Seasonal effect on certain developmental stages of *Tricholyga bombycis* Beck (Diptera, Tachinidae). *Proc. Indian Acad. Sci (Anim. Sci.).* **95**, 561-566.

Kumar Pradip and Jolly, M.S. and Venkatareddy, V., 1986. The effect of the number of maggots per host on the developmental stages of *Tricholyga bombycis* Beck, an endoparasite of *Bombyx mori* L. *Sericologia*, **26**, 519-528.

Kumar Pradip, Kumar, A. and Sengupta, K., 1990a. Parasitoids of uzi fly, *Exorista sorbillans* Wiedemann (Diptera, Tachinidae), IX. Effect of host and parasitoid age on parasitization and progeny production of *Nesolynx thymus*. *Indian J. Seric.*, **29**, 208-212.

Kumar Pradip, R. Kishore, and Sengupta, K., 1990b. Studies on the alternate hosts of uzi fly, *Exorista sorbillans* Wiedemann (Diptera, Tachinidae). *Indian J. Seric.*, **29**, 193-199.

Lingren, P.D. and Noble, L.W., 1972. Preference of *Campoletis perdistinctus* for certain noctuid larvae. *J. Econ. Entomol.*, **65**, 104-107.

Lingren, P.D., Guerra, R.J., Nickelson, J.W. and White, C., 1970. Hosts and host age preference of *Campoletis perdistinctus* (Viereck). *J. Econ. Entomol.*, **63**, 518-522.

Loke, W.H. and Ashley, T.R., 1984. Sources of fall armyworm, *Spodoptera fruigiperda* (Lepidoptera: Noctuidae), kairomones eliciting host finding behaviour in *Cotesia (=Apanteles) marginiventris* (Hymenoptera, Braconidae). *J. Chem. Ecol.,* **10,** 1019-1027.

Lotka, A.J., 1925. *Elements of physical biology*. Williams and Wilkins, Baltimore, Md. 462.

Mahadevappa, D., 1992. Uzi fly infestation and silkworm crop loss (Abstract). *Recent Advances in Uzi fly Research*. Proceedings of National Seminar on Uzi fly and its control. Eds. G.P. Channa Basavanna, G. Veeranna and S.B. Dandin, 20.

Manjunath, D. Kishore, R., Satya Prasad, K. Kumar, P. and Datta R.K., 1992. Biology of mulberry mealy bug and predatory potential of its biocontrol agent. National Conference on *Mulberry Sericulture Research,* 10th December at CSRTI, Mysore, 50.

Manjunath, H.B. and H.P. Puttaraju, 1996. Stages of oogenesis correlated with the age of the uzifly *Exorista bombycis (E. sorbillans)* (Diptera, Tachinidae). *Sericologia,* **36,** 107-118.

Monteith, L.G., 1955. Host preference of *Drino bohemica* Mesn. (Diptera, Tachinidae) with particular reference to olfactory responses. *Can. Ent.* **87,** 509-530.

Monteith, L.G., 1964. Influence of the health of the food plant of the host finding by tachind parasites. *Can. Ent.,* **96,** 1477-1482.

Monteith, L.G., 1967. Respones by *Diprion hercyniae* (Hymenoptera, Diprionidae) to its food plant and their influence on its relationship with its parasitoid *Drino bohemica* (Diptera, Tachinidae). *Can. Ent.,* **99,** 682-685.

Morris, R.F., Miller, C.S., 1954. Development of life tables for the spruce budworm. *Can. J. Zool.,* **32,** 283-301.

Murty, S.V., Jyothi, K.N. and K, Jamil, 1988. Environmental factors governing the infestation of uzi fly *Exorista sorbillans,* an endoparasite of mulberry silkworm *Bombyx mori* in certain areas of West Bengal and Andhra Pradesh. *Proc. Nat. Sym. on Nutritional Ecology of Insects and Environment,* Muzzaffarnagar, 74.

Nair, N.V., 1988. Sericulture - its place in national economy, *Indian Silk,* **6,** 22-24.

Nakata, M., 1963. Notes on the natural enemies of larvae and pupae of *Pieris melete* Meneties injurious to the Japanese horse radish. *Jap. J. appl. Ent. Zool.,* **7,** 348-349.

Narayanswamy, K.C., Devaiah, M.C. and R. Govindan, 1993. Studies on life table of uzi fly, *Exorista bombycis* (Diptera, Tachinidae). *Recent Advances in Uzi fly Research,* proceedings of National Seminar on uzi fly and its control. Eds. G.P. Channa Basavanna, G. Veeranna and S.B. Dandin, 31-42.

Narshimha Rao, Ch., Veeranna, G. and Prasad, N.R., 1993. Studies on the infestation of *Exorista sorbillans* on *Bombyx mori* L. incidence during monsoon season. *Recent Advances in Uzi Fly Research, Proceedings of National Seminar on uzi fly and its control.* Eds. G.P. Channa Basavanna, G. Veeranna and S.B. Dandin, 49-56.

Ngi-song, Adele J. and William A.O., 1997. Host location and acceptance by *Cotesia flavipes* Cameron and *C. sesamiae* (Cameron) (Hymenoptera, Braconidae), parasitoids of african gramineceous stemborers, Role of frass and other host cues. *Biological Control,* **9,** 136-142.

Nikam, P.K. and Sathe, T.V., 1983a. Studies on host age selection by *Cotesia flavipes* (Cameron), a larval parasitoid of *Chilo partellus* (Swin.). *Indian J. Parasitol.*, **7**, 181-182.

Nikam, P.K. and Sathe, T.V., 1983b. Life tables and intrinsic rate of natural increase of *Cotesia flavipes* (Cameron) (Hymenoptera, Braconidae) Population on *Chilo partellus* (Swin.) (Lepidoptera, Pyralidae). *Z. ang. Ent.*, **95**, 171-175.

Oatman, E.R. and Plantner, G.R., 1974. The Biology of *Temelucha* sp., *Platensis* group. (Hymenoptera, Ichneumonidae), a primary parasite of the potato tuberworm. *Ann. ent. Soc. Am.*, **67**, 275-280.

Odebiyi, J.A. and Oatman, E.R., 1972. Biology of *Agathis gibbosa* (Hymenoptera, Braconidae), a primary parasite of the potato tuberworm. *Ann. ent. Soc. Am.*, **65**, 1104-1114.

Osmolousky, G.E., 1964. On the biology of *Apanteles glomeratus* L. (Hymenoptera, Braconidae), a parasite of *Pieris brassicae* L. *Ent. Obozer*, **43**, 755-759.

Paroda, R.S., 1993. Crop science research for sustainable agriculture in India. *Pro. Nat. Acad. Sci. India*, **63**, 97-114.

Patil, G.M. and Govindan, R., 1984. Progeny production in uzi fly (in Kannada). *Reshme Krishi*, **6**, 17-20.

Persoons, C.J., 1988. Control of the uzi fly by semiochemicals. *Progress Report*, May-October, 1988 (R 89/035).

Persoons, C.J., 1989. Control of the uzi fly by semiochemicals. *Progress Report*, August, 1989-February, 1990 (R 89/033).

Persoons, C.J., 1990a. Control of the uzi fly by semoichemicals. *Progress Report*, August, 1989-February, 1990 (R 90/199).

Persoons, C.J., 1990b. Control of the uzi fly by semiochemicals. *Progress Report,* March September, 1990 (R 90/422).

Persoons, C.J., Veeranna, G., Vos, J.D. and Nagasundara, K.R., 1993. The role of semiochemicals in the host finding behaviour of parasitoids with speical reference to the uzi fly, *Exorista sorbillans* (Tachinidae). *Recent Advances in Uzi fly Research, Proceedings of National Seminar on uzi fly and its control.* Eds. G.P. Channa Basavanna, G. Veeranna and S.B. Dandin, 167-189.

Prasad, N.R. Reddy, Y.S. and Krishnamurty, N.B., 1985. Studies on the physiological interactions between silkworm (*Bombyx mori*) and its endoparasite uzi fly *Exorista sorbillans. Sericologia,* **25,** 511-523.

Rajshekhargouda, R. and Devaiah, M.C., 1983. Occurrence of uzi fly, *Exorista sorbillans* Wiedemann (Diptera, Tachinidae) in northern districts of Karnataka (Abstract), *Natl. Semi. Silk Res. Dev.,* Bangalore, 113.

Raju, P.J., Mahadevappa, R., Raghuraman and S.B. Dandin, 1995. Sericulture among tribals of B.R. Hills. *Indian Silk,* **34,** 10-12.

Ramakant, Y.R., Madhava Rao and A. Sarkar, 1998. Soil moisture management in mulberry. *Indian Silk,* **36,** 5-7.

Ram Kishore, Pradip Kumar, D. Manjunath, 1993. Biology of *Tetrastichus howardi,* a parasitoid of uzi fly, *E. sorbillans. Recent Advances in Uzi Fly Research, Proceedings of National Seminar on uzi fly and its control.* Eds. G.P. Channa Basavanna, G. Veeranna & S.B. Dandin, 135-142.

Rao, U.R., 1996. Presidential address. *83rd Indian Science Congress, Patiala,* 1-48.

Rath, B.B., 1966. Fly parasites of tasar silkworm, *Indian Silk,* **5,** 8.

Ray, Indrajit, 1989. Can mulberry provide an answer to the scourage of AIDS. *Indian Silk,* **27,** 37-38.

Reddy, D.N.R. and Kotikal, Y.K., 1988. Pests of mulberry and their management. *Indian Silk,* **3,** 9-14.

Roland, J., Evans, W.G. and Myers, J.H., 1989. Manipulation of oviposition patterns of the parasitoid *Cyzenis albicans* (Tachinidae) in the field using plant extracts. *J. Ins. Behavi.,* **2,** 487-503.

Roth, J.R., King, E.G. and Thompson, A.C., 1978. Host location behaviour by the tachinid *Lixophaga diatraea. Environ. Entomol.,* **7,** 794-798.

Sabrosky, C.W. and Reardon, R.C., 1976. Tachinid, parasites of the gypsy moth *Lymantria dispar,* with key to adults and Puparia. *Entomol. Soc. Am.,* **10,** 55-57.

Sanan, J., 1934. *Trichology sorbillans* Wied. *Formo San. Agr. Rev.,* **30,** 666-667.

Salt, G., 1935. Experimental studies in insect parasitism. III-Host-selection. *Proc. Roy. Soc. Ser. B. Biol. Sci.,* **117,** 413-435. D.C., 1980. Sericulture in India. CSB, Bombay, 51.

Sarkar, D.C. 1980. Sericulture in India, CSB, Bombay, 51.

Sathe, T.V., 1986a. Biology of *Cotesia diurnii* Rao and Nikam (Hymenoptera, Braconidae), a larval parasitoid of *Exelastis atomosa* Walsingham. *Oikoassay,* **3,** 32-33.

Sathe, T.V., 1986b. Life tables and intrinsic rate of increase of *Cotesia diurnii,* Rao and Nikam (Hymenoptera, Braconidae), a larval parasitoid of *Exelastis atomosa,* Wals. *Entomon,* **11,** 281-283.

Sathe, T.V., 1987a. Morphology and biometry of immature stages of *Cotesia orientalis* C. & N. (Hymenoptera, Braconidae), an internal larval parasitoid of *Exelastis atomosa* Wals. *Uttar Pradesh J. Zool.* **7,** 200-203.

Sathe, T.V., 1987b. Life tables and intrinsic rate of natural increase of *Diadegma trichoptilus* (Cameron) (Hymenoptera, Ichneumonidae) population on *Exelastis atomosa* Wals. (Lepidoptera, Pterophoridae) *J. Adv. Zool.*, **8,** 1-4.

Sathe, T.V., 1988a. Biology of *Diadegma trichoptilus* (Cameron) (Hymenoptera, Ichneumonidae), a larval parasitoid of *Exelastis atomosa* Walsingham. *J. Curr. Biosci.*, **5,** 37-40.

Sathe, T.V., 1988b. The biology of *Cotesia orientalis* Chalikwar and Nikam (Hymenoptera, Braconidae), a larval parasitoid of plume moth in India. *J. Zool. Res.* **1,** 23-27.

Sathe, T.V., 1988c. Intrinsic rate of increase and interspecific relationship between *Cotesia orientalis* C. & N., *C. diurnii* R. & N. and *Diadegma trichoptillus* (Cameron), the larval parasitoids of *Exelastis atomosa* Walsingham. *Advances in Parasitic Hymenptera Research,* U.S.A., 383-387.

Sathe, T.V., 1990a. The biology of *Diadegma argenteopilosa* Cameron (Hymenoptera, Ichneumonidae), an internal larval parasitoid of *Spodoptera litura* (Fab.). *The Entomoligist,* **109,** 2-7.

Sathe, T.V., 1990c. Life table of *Parenion bhairavi* population on *Diacrisia obliqua,* (Wlk.). *Uttar Pradesh J.Zool,* **11,** 162-166.

Sathe, T.V., 1992. Natural enemies of some insect pests of economic importance. *Oikoassay,* **9,** 15-17.

Sathe, T.V., 1993. Host age selection by *Apanteles jayanagarensis* Bhatnagar (Hymenoptera, Braconidae), a parasitiod of *Spilosoma obliqua* (Wlk.) *Comp. Physiol.Ecol.* 18, 140-143.

Sathe, T.V., 1998. *Sericultural crop protection*. Asawari Publications. 1-130.

Sathe, T.V. and D.M. Ingawale, 1993a. Host age selection by *Apanteles blateatae* (Hymenoptera, Braconidae), a parasitoid of *Sylepta derogata*. *Trends in Life Sciences*, **8**, 113-116.

Sathe, T.V., and D.M. Ingawale, 1993b. Fecundity, life table and Intrinsic rate of increase in *Apanteles jayanagarensis* Bhatnagar, a larval parasitoid of *Spilosoma obliqua*, (Wlk.) *J. Biol. Control.*, **7**, 69-71.

Sathe, T.V., and Mulla, M.K., 1999. Insect pests of mulbery from Kolhapur, India. *Geobios new Reports*, **18**, 73-74.

Sathe, T.V. and Nikam, P.K., 1986. Life tables and intrinsic rate of increase of *Campoletis chlorideae* (Hymenoptera, Ichneumonidae), an internal larval parasitoid of *Heliothis armigera* (Hubn.). *Rev. di Parassitologia*, **3**, 289-294.

Sathe, T.V., and Santha Kumar, M.V., 1988a. Factors responsible for host finding behaviour by *Campoletis chlorideae* Uchida (Hymenoptera, Ichneumomidae), a parasitoid of *Heliothis armigera* (Hubner) (Lepidoptera, Noctuidae). *Rev. di parassitologia*, **5**, 233-240.

Sathe, T.V. and Santha Kumar, M.V., 1992a. Host specificity in *Campoletis chlorideae* Uchida (Hymenoptera, Ichneumonidae). *J. Adv. Zool.* **13**, 53-56.

Sathe, T.V., Santha Kumar, M.V. and S.A. Inamdar, 1988. The biology of *Apanteles creatoniti* Viereck. (Hymenoptera, Braconidae), a larval parasitoid of *Thiocidas postica* Wlk. (Lepidoptera). *Entomon* **13**, 189-190.

Sathe, T.V., Mulla, M.K., Arballi, Gajendra and Rupali Pawar, 1999. White fly - A new pest of mulberry. *Indian Silk*, **37**, 10.

Sathe, T.V., Santha Kumar, M.V., Indamdar, S.A. and Ingawale, 1987a. Reproductive potential of *Apanteles creatonoti* Viereck (Hymenoptera) in relation to age of *Thiocidas postica* (Wlk.) caterpillar (Lepidoptera) *Uttar Pradesh. J. Zool.* **1,** 89-91.

Sathe, T.V., Nikam, P.K., Jadhav, R.V. and Phadtare, S.T. 1987b. Influence of age of caterpillers of *Exelastis atomosa* Wals. (Lepidoptera, Pterophoridae) on parasitization by *Cotesia orientalis. Indian J. Parasitol.,* **11,** 87-88.

Sato, Y., 1979. Experimental studies on parasitization by *Apanteles glomeratus,* IV. Factors leading a female to the host. *Physiol. Entomol.,* **4,** 63-70.

Savanurmath, C.J. and Patil, G.M., 1993. Biology and behaviour of adult Tasar uzi fly *B. zebina. Recent Advances in Uzi Fly Research, Proceedings of National Seminar on uzi fly and its control.* Eds. G.P. Channa Basavanna, G. Veeranna and S.B. Dandin, 257-267.

Sengupta, K., Kumar, P., Baig M. and Govindaiah, 1990. Handbook on diseases and pests of silkworm and mulberry. *United Nations, ESCAP,* Bankon, Thailand, 35.

Shetty, K.K. and Samson, M.V., 1998. Non-mulberry sericulture in India. *Indian Silk,* **37,** 21-25.

Siddappaji, C., 1985. Bioecology and management of the Indian uzi fly *Exorista sorbillans* (Diptera, Tachinidae), a parasite of mulberry silkworm. *Ph. D. Thesis, U.A.S. G.K.V.K.,* Bangalore, 126.

Siddappaji, C. and Basavanna G.P.C., 1990. The Indian uzi fly, *Exorista bombycis,* a parasitoid of mulberry silkworm. *Indian J. Seri.,* **29,** 119-137.

Siddappaji, C. and Kaiser Jamil, 1992. Host range of the Indian uzi fly, *Exorista biombycis* (Abstract). *Natl. Semi. on uzi fly and its control KSSDI,* Bangalore, 11.

Siddiqui, A.A., 1988. Tasar culture and its role in community development. *Indian Silk,* **26,** 35-37.

Siddiqui, A.A. and Omkar, 1987. Impact of B.S.M. & T. on tribal upliftment. *Indian Silk,* **25,** 21-22.

Sinchaisir, N., Isarangkul, L. and Ayuthaya, N., 1972. The host preference of the tachinid fly to different stages and varieties of silkworm. *Bull. Thai. Sericult. Res. Train. Centre.,* **2,** 67-69.

Singh, A. and Sadhu, A.N., 1986. *Agricultural problems in India.* Jammu University, Jammu, 386.

Singh, K.P. and Das, P.K., 1996. Some records of pests and predators of primary muga food plants and silkworm *A. assama* W. in Assam. *Sericologia,* **36,** 763-765.

Singh, K.P. and Ghosh, P.L., 1992. Mulberry cultivation under agro forestry and land management. *Indian Silk,* **31,** 16-18.

Slobodkin, L.B., 1962. *Growth and regulation of Animal Populations.* Holt Reinhart and Winston, New York, 184.

Smilowitz, Z., 1974. Relationship between the parasitoid *Hyposoter exigua* (Viereck) and cabbage looper, *Trichopulsia ni* (Hubner) : Evidence for endocrine involvement in successful parasitism. *Ann. ent. Soc. Am.,* **67,** 317-320.

Smilowitz, Z. and Iwantsch. G.E., 1975. Relationships between the parasitoid *Hypostor exigua* and the cabbage looper, *Trichoplusia ni*. The effect of host age on ovipositional rate of the parasitoid and successful parasitism. *Can. Ent.* **107,** 689-694.

Sood, A.K., Bhalla, O.P., Sharma, K.C. and Anil Kumar, 1995. Seasonal activity of natural enemies of *Pieris brassicae* (Linnaeus) (Lepidoptera, Pieridae) in cauliflower seed crop ecosytem. *J. Biol. Control,* **9,** 119-122.

Spradberry, J.P., 1968. A technique for artificially culturing ichneumonid parasites of wood wasp. *Ent. exp. & Appl.,* **11,** 257-260.

Sree Kumar, S., Nair Sudhakaran, G., Appaswamy, P., Vijayaraghavan, K. and V. Thiagarajan, 1994. Now, mulberry on your dining table. *Indian Silk,* **32,** 45.

Sriharan, T.P., Samson, M.V., Krishnaswami, S. and Datta, R.K., 1971. Laboratory investigations on uzi fly, *Tricholyga bombycis* Beck., a tachinid parasite of silkworm (*Bombyx mori* L.) : Rearing and reproduction of the parasite. *Indian J. Seric.,* **10,** 4-22.

Sufty, R.K. and E. Fuhrer, 1981. Parasitare veranerungen der wirtskutikula bei *Pieris brassicae* and *Cydia pomonella* Durch. entompmophage endoparasiten. *Ent. exp. & appl.,* **30,** 134-139.

Sukhatme, P.V., Sukhatme, B.V. and C. Asoke, 1984. *Sampling theory of survey with application*. Lowa state Univ. press and ISAS, New Delhi, 526.

Swaminathan, M.S., 1994. Biotechnology and fighting the fatigue of green revloution. *Proc. II Asia Pacific conference on agricultural biotechnology,* Madras.

Tagawa, J., 1984. Diapause in the braconid wasp *Apanteles glomeratus L.* (I) Evidence of diause in over wintering prepupae. *Appl. Ent. Zool.,* **19,** 396-399.

Tanaka, Y., 1964. *Sericology*. CSB, Bombay, 177.

Tawfik, M.F.S., 1957. Host parasite specificity in a braconid *Apanteles glomeratus* L. *Nature,* **179,** 1031-1032.

Tayade, D.S., 1991. Sericulture in Maharashtra : Challenges and Solutions. *Indian Silk,* 30-32.

Tewary, P.K. and G. Subha Rao, 1990. Mulberry Bi-products and their utilization. *Indian Silk,* **29,** 43-44.

Thangavelu, K. 1993. Natural enemies of muga silkworms, *Antheraea assama* West wood (Lepidoptera, Saturniidae). *J. Insect Sci,* **6,** 57-59.

Thangavelu, K., and R.N. Singh 1993. Integrated management of uzi fly *Blepharipa zebina* in tasar ecosystem. *Recent Advances in Uzi fly Research, Proceedings of National Seminar on uzi fly and its control.* Eds. G.P. Channa Basavanna, G. Veeranna and S.B. Dandin.

Thomas, Jacob, 1998. Sericulture Development in Uttar Pradesh. *Indian Silk,* **37,** 5-9.

Thompson, W.R., 1944. A catalogue of the parasites and predators of insect pests. Sect. I, Part - 5. *Imperial Agricultural Bureau,* Canada, 130.

Thompson, W.R., 1951. A catalogue of the parasites and predators of insect pests. Sec. II, Part - I. *Commonwealth Agr. Bull.,* Ottawa, Canada. 14.

Thompson, W.R., 1953. *A catalogue of the parasites and predators of insect pest*. Section II, Host of the Hymenoptera, Agathidae to Braconidae. 64-65.

Thompson, W.R., 1954. A catalogue of the parasites and predators of insect pests. Section I, Parasite host catalogue part 5. *Parasites of the Lepidoptera,* 98.

Thrope, W.H., 1932. Experiments upon respiration in the larvae of certain parasitic Hymenoptera. *Proc. Roy. Soc. Lond. Ser. B.,* **109,** 450-471.

Thurston, R. and Postely, L., 1978. Effect of instar of *Manduca sexta* on the rate of development of its parasite, *Apanteles congregatus. Tob. Sci.,* **22,** 32-34.

Vender Wulp, F.M., 1986. Parasitic muscidae from British India - 3, *Tricholyga bombycis* Becher. *Indian Museum. Notes.* **3,** 11-15.

Venkatadri, J.S., 1980. Incidence of uzi fly (in Kannada). *Reshme Krishi.,* **2,** 29-31.

Venugopal, V., 1994. Women in sustainable sericulture development. *Indian Silk,* **33,** 29-32.

Vijayan, K., S.P. Chakraborti, B.N. Roy and S.K. Sen, 1998. West Bengal : Winter Hardy Mulberry variety: A need. *Indian Silk,* **37,** 6-8.

Vinson, S.B., 1965. A method of host selection by *Cardiochiles nigriceps. J. Econ. Entomol.* **58,** 869-71.

Vinson, S.B., 1976. Host selection by insect parasitoids. *A. Rev. Ent.,* **21,** 109-134.

Vinson, S.B., 1980. Habitat location in semiochemicals, their role in pest control, 51-78. D.A. Nordlund, R.L. Jones, and W.J. Lewis, Ed. J. Wiley & Sons, New York, U.S.A.

Waage, J.K. and Greathead, D., 1986. Insect Parasitoids. Academic Press London, 1-375.

Watanabe, C., 1964. A tentative catalogue of insect natural enemies of injurious insects in Japan. Part I. Parasite - predator host catalogue (compiled by Yasumatsu, K. and C. Watanabe). *Ent. Lab. Fac. Agric. Kyushu Univ.,* 61.

Weseloh, R.M., 1980. Host recognition behaviour of the tachinid parasitoid, *Compsilura concinnata. Ann. ent. Soc. Am.,* **73,** 593-601.

Weseloh, R.M., 1981. Host location by parasitoids. In: *Semiochemicals their role in pest control* (D.A. Nordlund, R.L., Jones and W.J. Lewis (eds.), 79-95.

Yadav, G.S., Reddy, K.J., Murgod, S.B., Roy, G.C. and B.R.R.P. Sinha, 1997. Medicinal Tasar flora of Bhandara. *Indian Silk,* **36,** 34-37.

Yazgan, S. and House, H.L. 1970. A hymenopterous insect, the parasitoid *Itoplectis conquisitor,* reared axiematically on a chemically defined synthetic diet. *Can. Ent.* **102,** 1304-1306.

SUBJECT INDEX

SPECIES INDEX